Should We Eat Meat?

Should We Eat Meat?

Evolution and Consequences of Modern Carnivory

Vaclav Smil

A John Wiley & Sons, Ltd., Publication

This edition first published 2013 © 2013 by John Wiley & Sons, Ltd

Wiley-Blackwell is an imprint of John Wiley & Sons, formed by the merger of Wiley's global Scientific, Technical and Medical business with Blackwell Publishing.

Registered Office
John Wiley & Sons, Ltd, The Atrium, Southern Gate, Chichester, West Sussex, PO19 8SQ, UK

Editorial Offices
9600 Garsington Road, Oxford, OX4 2DQ, UK
The Atrium, Southern Gate, Chichester, West Sussex, PO19 8SQ, UK
2121 State Avenue, Ames, Iowa 50014-8300, USA

For details of our global editorial offices, for customer services and for information about how to apply for permission to reuse the copyright material in this book please see our website at www.wiley.com/wiley-blackwell.

Library of Congress Cataloging-in-Publication Data

Smil, Vaclav.
   Should we eat meat? : evolution and consequences of modern carnivory / Vaclav Smil.
      pages   cm
   Includes bibliographical references and index.
   ISBN 978-1-118-27872-7 (softback : alk. paper) – ISBN 978-1-118-27868-0 (emobi) – ISBN 978-1-118-27869-7 (epub) – ISBN 978-1-118-27870-3 (epdf) – ISBN 978-1-118-27871-0   1. Meat–Health aspects.
2. Meat industry and trade.   3. Vegetarianism.   I. Title.
   QP144.M43S65 2013
   641.3′6–dc23
                                                        2013001800
A catalogue record for this book is available from the British Library.

Wiley also publishes its books in a variety of electronic formats. Some content that appears in print may not be available in electronic books.

Cover images: Meat image: © iStockphoto/mnicholas; Dear image © iStockphoto/DamianKuzdak; Cow image © iStockphoto/meadowmouse
Cover design by Meaden Creative

Set in 10/12.5pt Galliard by SPi Publisher Services, Pondicherry, India
Printed and bound in Malaysia by Vivar Printing Sdn Bhd

2 2014

About the author

Vaclav Smil conducts interdisciplinary research in the fields of energy, environmental and population change, food production and nutrition, technical innovation, risk assessment and public policy. He has published more than 30 books and close to 500 papers on these topics. He is a Distinguished Professor Emeritus at the University of Manitoba, a Fellow of the Royal Society of Canada (Science Academy), the first non-American to receive the American Association for the Advancement of Science Award for Public Understanding of Science and Technology, and in 2010 he was listed by *Foreign Policy* among the top 50 global thinkers.

Other books by this author

China's Energy
Energy in the Developing World
  (edited with W. Knowland)
Energy Analysis in Agriculture
  (with P. Nachman and
  T. V. Long II)
Biomass Energies
The Bad Earth
Carbon Nitrogen Sulfur
Energy Food Environment
Energy in China's Modernization
General Energetics
China's Environmental Crisis
Global Ecology
Energy in World History
Cycles of Life
Energies
Feeding the World

Enriching the Earth
The Earth's Biosphere
Energy at the Crossroads
China's Past, China's Future
Creating the 20th Century
Transforming the 20th Century
Energy: A Beginner's Guide
Oil: A Beginner's Guide
Energy in Nature and Society
Global Catastrophes and Trends
Why America Is Not a New Rome
Energy Transitions
Energy Myths and Realities
Prime Movers of Globalization
Japan's Dietary Transition and
  Its Impacts (with K. Kobayashi)
Harvesting the Biosphere

Contents

Pieter Brueghel the Elder filled his rich kitchen with obese diners devouring suckling pigs, hams and sausages. Detail of the engraving produced by Hans Liefrinck in 1563.

Preface

Carnivory of modern Western societies is constantly on display. Supermarkets have meat counters that are sometimes tens of meters long, full of scores of different cuts (or entire eviscerated carcasses) of at least half a dozen mammalian and avian species (cattle, pig, sheep, chicken, turkey, duck). Some of them, and many specialty shops, also carry bison, goat and ostrich meat, as well as pheasants, rabbit and venison. Then there are extensive delicatessen sections with an enormous variety of processed meat products. Fast-food outlets – dominated by ubiquitous burger chains – were built on meat, and despite their recent diversification into seafood and vegetarian offerings, they remain based on beef and chicken. Consumption statistics confirm this all too obvious extent of carnivory, with annual per capita supply of meat at retail level (including bones and trimmable fat) surpassing typical adult body weights (65–80 kg) not only in the US and Canada and in the richer northern EU nations but now also in Spain. In fact, Spanish per capita meat supply has been recently the Europe's highest.

What all but a few typical (i.e., urban) carnivores do not realize is the extent to which the modern Western agriculture turns around (a better way to express this would be to say: is subservient to) animals: both in terms of the total cultivated area and overall crop output, it produces mostly animal feed (dominated by corn and soybeans) rather than food for direct human consumption (staple grains dominated by wheat, tubers, oilseeds, vegetables). But, if they are so inclined, modern Western urbanites can find plenty of information about the obverse of their carnivory, about poor treatment of animals, about environmental degradation and pollution attributable to meat production, and about possible health impacts. Vegetarianism has been an increasingly common (but in absolute terms it is still very restricted) choice among the Western populations, and vegetarian publications and websites have been a leading source of information on the negatives of carnivory. Vegans in particular enumerate the assorted sins of meat eating in an often strident manner on many Internet sites. These contrasting attitudes have been reflected in the published record.

On one hand, there are hundreds of meat cookbooks – unabashed and colorfully illustrated celebrations of meat eating ranging from several "bibles" that are devoted to meat in general (Lobel et al. 2009; Clark and Spaull 2010) and to meatballs and ribs in particular (Brown 2009; Raichlen 2012) or to *Grilling Gone Wild* (Couch 2012) – all promising the best-ever, classic, succulent, complete meat repasts. The middle ground of meaty examinations is occupied by what I would call mission books of many gradations, the mildest ones imploring their readers to eat less meat (Boyle 2012) or arguing the benefits of becoming a flexitarian, that is, only an occasional meat eater (Berley and Singer 2007). The more ambitious ones are trying to convert meat eaters into vegetarians, even vegans, in ways ranging from straightforward (de Rossi et al. 2012) to enticing (O'Donnel 2010). And the same contrasts and arguments have been replayed in yet another genre of books that examine meat's roles in national and global history (Rimas and Fraser 2008; Ogle 2011).

Finally, there is a venerable tradition of books as instruments of indictment. This genre began in 1906 with Upton Sinclair's novel uncovering the grim realities of Chicago's stockyards and meatpacking (Sinclair 1906). A reader entirely unfamiliar with the revolting nature of Sinclair's descriptions will find an extended quote in a section about meat processing. More than a century later, critics of anything associated with meat include such disparate groups as activists agitating for animal rights, environmental scientists worried about cattle taking over the planet and nutritionists convinced (not quite in accord with the complete evidence) that eating meat undermines health and hastens the arrival of death.

Some of these writings portray modern meat industry in truly gruesome terms, and many have unsubtle titles or subtitles that make it clear that meat production and animal slaughtering are components of a despicable, if not outright criminal, enterprise and that meat eating is a reprehensible habit, a deplorable ride that must end: meat is madness (Britton 1999); meat animals are devouring a hungry planet (Tansey and D'Silva 1999); meat production is a matter of crimes unseen (Jones 2004); and eating meat is our society's greatest addiction (Ford 2012). Others, including books by Schlosser (2001) and Pollan (2006), are more measured in their condemnation. But in terms of extreme positions and incendiary language, few texts can beat *The Sexual Politics of Meat* by Carol Adams first published in 1990: the book's subtitle claims to offer a feminist-vegetarian critical theory; it abounds in such deliberately provocative phrases as "the rape of animals" and "the butchering of women."

And this is how it ends: "Eat Rice Have Faith in Women. Our dietary choices reflect and reinforce our cosmology, our politics. It is as though we could say, 'Eating rice *is* faith in women.' On this grace may we all feed" (Adams 2010, 202). Of course, the nation where rice had a more prominent place in social identity, self-perception and culture than anywhere else – and where the plant has been a cherished symbol of wealth, power and beauty (Ohnuki-Tierney 1993) – would be left entirely unmoved by that argument: average per capita consumption of rice in Japan is now, in mass terms, lower than the intake of dairy products (less than 65 vs. more than 80 kg/year), and the country's accurate food balance sheets show the per capita supply of meat and seafood at nearly 100 kg/year, or 50% higher than rice (Smil and Kobayashi 2012). And Japan has been a model followed by other traditionally rice-eating Asian nations whose rice consumption appears to fall by about a third with every doubling of income (Smil 2005a).

What are we to make of these contending and contradictory conclusions? Should we eat meat – or should we try to minimize its consumption and aim at its eventual eliminations from human diets? My answers will be based on long-term perspectives and on complex and multidisciplinary considerations: my appraisals of the evolution of meat eating, historical changes and modern modalities of this practice and its benefits as well as its undesirable consequences are based on findings from disciplines ranging from archaeology to animal science and from evolutionary biology to environmental and economic studies. This is a book rooted in facts and realities, not in predetermined posturing and sermonizing, a book that looks at benefits of meat eating as well as at the failures and drawbacks, and that does not aspire to fit into any pre-cast categories, pro or contra, positively programmatic or aggressively negative. I do not approach the reality of modern large-scale carnivory with any pre-conceived notions, and I did not write this book in order to advocate any particular practice or point of view but merely in order to follow the best evidence to its logical conclusions. At its end, a reader will know quite clearly where I stand – but I thought that at its beginning it might be interesting to explain where I come from, that is, to make my full meat-eating disclosure.

As a child, adolescent and a young man I ate a wide variety of meat, but never in large individual portions or in large cumulative quantities: realities of post-WW II Europe (in some countries food rationing was in place until 1954), my mother's cooking and my food preferences (I have always disliked large and thick pieces of meat and all fatty cuts) explain that. But this moderation went along with a great variety, and before I left Europe for North America at the age of 25, I had eaten pork, beef, veal, mutton, lamb, goat, horse, rabbit, chicken, duck, pigeon, goose, turkey and pearl hen, and as a boy the meat I loved best came from the animals my father shot during the hunting season, pheasants, wild hares and, above all, deer.

Also as a child I attended with my parents a number of village winter pig killings at the houses of my father's acquaintances. In many traditional European societies, these used to be (and in some places still remain) festive social and culinary events: *Schlachtfest* in Germany, *maialata* in Italy, *matanza* in Spain and *zabíjačka* in Bohemia. They are crowned by eating a remarkable variety of foodstuffs prepared expertly from the killed animal – including blood soup, blood sausages, white sausages and headcheese – and the attendants then take home assorted lean and fat cuts to be roasted or boiled or processed into lard. Other meat-related memories of my childhood include: my grandmother force-feeding geese (a practice I disliked); my father placing fragrant evergreen boughs into the cavity of deer carcass before hanging it to age in cold air (so learning as a child that fresh meat is not really fresh); my mother cooking beef *rouladen* stuffed with carrots, onions, boiled eggs and gherkins (yielding a colorful combination of fillings that is beautifully revealed on cooked cross-cut).

And as in any traditional society, when I was a child we also ate organ meats, albeit much less frequently than pork or chicken. Except for tripe (a preference I share with all those who like *trippa alla romana*) they were never my favorites, but I had eaten brains, lungs, heart, kidneys, cow's udder and calf, pig and poultry livers, the latter both cooked and prepared as *pâtés*. Quantifying those childhood and adolescent meat intakes is impossible with accuracy, but my best estimate is that as a teenager my annual meat consumption was on the order of 15 kg, with a few more kilograms of processed meats (mainly ham and sausages). Liver was the only organ meat whose eating had temporarily survived our move across the Atlantic: for a time during the 1970s, I used to make fairly regularly a chicken liver *pâté* with cognac.

But our trans-Atlantic move and the access to much cheaper meat in general, and to inexpensive beef in particular, did not change my dislike of large or fatty pieces of meat: as a result, in more than four decades of living (and almost daily cooking) in North America, I have never eaten, bought or cooked a steak – and, a fact many readers might find even harder to believe, I have never eaten a hamburger in McDonald's or in any of America's other burger chains. Virtually all cooking with meat I did during the years when our son was growing up was Chinese and Indian food, with small pieces of meat in sauces and with vegetables served with rice; the only exceptions were some holiday roasts and an occasional *Wiener Schnitzel.* During that period (in order to understand better the difficulties involved in performing representative food intake surveys), I had repeatedly monitored our actual food intakes for a few days at a time, and so I can state with certainty that our average annual per capita meat consumption had never surpassed 25 kg.

When our son left for graduate school in 1996, we continued to eat all animal foods (especially fish, cheeses and yogurt) but cooked red meat and poultry only a few times a year for traditional holiday dinners. According to an inaccurate, cumbersome but often used current dietary parlance, we became the longest-definition vegetarians (lacto-ovo-pisci-vegetarians). There were no sudden specific reasons behind this shift, just a naturally evolving preference. During that time, our per capita meat intake was well below 5 kg/year, but after a dozen years of these virtually meatless diets I began cooking again occasional meat dishes, including my favorite Indian curries and *Schnitzel,* and occasionally buying some good-quality *prosciutto* or *jamón Serrano* – and a few days before these introductory lines were written, I ate in Firenze a small dish of *trippa alla fiorentina,* a cook's natural curiosity to try once again an ancient local favorite.

During the most recent years, our total per capita meat consumption (actually consumed servings, not retail weight, although with the lean meat and boneless cuts I buy these two categories are pretty close) has been thus less than 5 kg/year. As with most people in the West, I should thus be classified as a life-long omnivore – but one with an increasing tendency toward very low meat consumption. After finishing this book, some readers may find that my dietary preferences had some effects on its tenor and on its conclusions; as a scientist, I would like to think that has not been the case, but others may conclude differently.

With this confession out of the way, I am ready to plunge into the realities, complexities and consequences of modern meat production and consumption. In Chapter 1, I must lay out first assorted meat basics, many essential facts and observations about meat in nutrition and health: about its properties, composition, quality and variety; about its role in human diets, above all as a source of high-quality protein and some key micronutrients, and its association with fat; and about its demonstrated and suspected roles in the genesis of major civilizational diseases and in human longevity. Some of these fascinating, but often inconclusive, links between meat and health and longevity have received a great deal of research attention, but they are also subject to an even greater amount of false beliefs and misinformation, and I will try to sort out this complex relationship by referring to the best available evidence. A separate section will be devoted to diseased meat and impacts and risks of meat-borne pathogens.

In Chapter 2, I will explain the evolutionary basis of human diets and their historical development extending from domestication of animals to typical meat intakes in traditional societies and including dietary taboos and proscriptions as well as meat's common position as a prestige food. Chapter 3 opens with a brief review of modern dietary transition, a process that has transformed traditional diets and whose two main components have been reduced consumption of carbohydrate staples and higher intakes of animal foodstuffs in general and meat in particular. This will be followed by an introduction to modern meat production and consumption that will trace the meat chain from the reproduction and growth imperatives through slaughtering of animals and processing of meat to meat consumption and waste, and that will systematically sort out the statistical categories used to quantify and compare these processes in historical and international terms.

Chapter 4 will explain what it takes to produce meat: it will first survey the modalities of modern animal husbandry (grazing, mixed farming and centralized "landless" industrial systems) before turning to long-term changes and current best practices of efficient meat production using balanced feed rations and to (humane as well as abusive) treatment of meat animals. The chapter's second part will focus on the environmental consequences of modern mass-scale carnivory. This is not a new concern but one that has gained a much higher prominence as the attention to environmental degradation and pollution and the concerns about the state of the biosphere and the sustainability of modern civilization have become increasingly common subjects of scientific inquiry, public discourse and governmental policy.

Animal Agriculture and Global Food Supply, a comprehensive report prepared by the US Council for Agricultural Science and Technology, was the first notable contribution to these new, environmentally centered, perspectives on livestock and meat production (Bradford et al. 1999). While it detailed many concerns, it concluded that livestock have both positive and negative environmental effects. Seven years later, an even more substantial interdisciplinary report prepared by the FAO had a different message, giving away its concerns by its very title: *Livestock's Long Shadow* (Steinfeld et al. 2006).

Its basic conclusion made many headlines:

> The livestock sector emerges as one of the top two or three most significant contributors to the most serious environmental problems, at every scale

from local to global. The findings of this report suggest that it should be a major policy focus when dealing with problems of land degradation, climate change and air pollution, water shortage and water pollution and loss of biodiversity.

The report's most often cited findings were that 26% of the Earth's surface is devoted to grazing land, 33% of all arable land is used to grow feed for animals, 18% of all greenhouse gas emissions are attributable to livestock as is 8% of the total use of freshwater.

Two years after FAO's report came a study that had a narrower focus but whose conclusions were even more worrisome: the goal of the Pew Commission on Industrial Farm Animal Production (IFAP) was "to sound the alarms" as it determined that "the negative effects of the IFAP system are too great and the scientific evidence is too strong to ignore. Significant changes must be implemented and must start now. And while some areas of animal agriculture have recognized these threats and have taken action, it is clear that the industry has a long way to go" (PCIFAP 2008). All of these three reports are readily available on the Web, and I will refer to them only when I will need to make, or stress, some specific points that are well developed in these studies. Rather than repeating much they have to offer, I will question some of their approaches and conclusions in an effort to demonstrate many uncertainties that make some of the published conclusions much less definitive than they may appear when they are cited as absolute findings.

I will do this by surveying five major categories of impact. The first one is the 20th century's rapid rise of domesticated zoomass and its densities, a topic that has not been addressed by the three reports. The second concerns changing animal landscapes, with the effects on land cover and land use dominated by deforestation, deliberately set fires, grazing and overgrazing. Intensive production of feedstuffs is the main reason why meat is an expensive food in virtual energy terms, with indirect energy costs due to intensive cultivation of feedstuffs being far more important than direct energy costs of feeding, housing and killing animals and processing, distributing and cooking meat.

And the last two categories of environmental concern will deal with the aquatic and atmospheric impacts of meat production. Large volumes of virtual water are needed to grow animal feeds, and the nutrients lost during that process as well as copious metabolic by-products of meat production are major water pollutants that contribute to undesirably high nutrient loadings and eutrophication of both fresh and coastal ocean waters, while gases released during the cultivation of feed crops and by metabolizing animals are significant factors in local, regional and global atmospheric changes (the last instance being due to emissions of methane and nitrous oxide, two major greenhouse gases). Some of the published claims appear less dire when seen in a proper context, but there is no doubt that when compared with burdens imposed by other foodstuffs, meat has a high environmental cost.

I will conclude the book with an appeal for what I call rational meat eating. I will first assess the extent to which non-meat options – vegetarianism or diets enriched with other animal foodstuffs (including the promise of *in vitro* meat) – can displace current meat eating, and then I will outline a path of desirable meat production. Advocacy of such a path will anger vegans and it will disappoint vegetarians – while its insistence on moderation will not satisfy the proponents of unrestrained, vigorous carnivory. But I believe that such a choice offers the best way to preserve social, economic and nutritional benefits of meat eating while minimizing many unavoidable and undesirable environmental impacts of mass-scale meat production.

Should We Eat Meat?

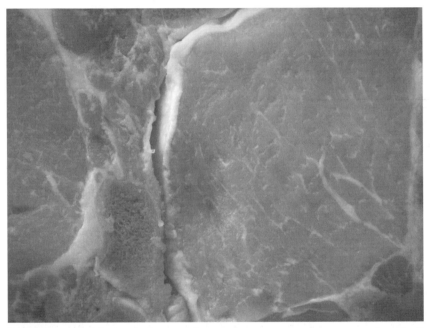

Pork loin center chops. A close-up shows what most meat cuts are composed of: muscle fascicles, collagen sheaths, tendons, intra- and extramuscular fat, and bones. Photo by V. Smil.

1

Meat in Nutrition

First things first: no energy conversion is more fundamental for the survival of our species than photosynthesis (primary productivity), the source – directly in raw or processed plants and indirectly in (usually cooked or processed) animal tissues – of all of our food. Eating (setting aside food smells, taste, visual appeal and all those cultural and historical connotations subsumed in the act of ingestion) can be defined in the most reductionist biophysical fashion as a process that supplies macronutrients (carbohydrates, proteins, lipids) and micronutrients (vitamins and minerals) that are required to sustain our metabolism needed for growth, maintenance and activity and hence to perpetuate life of this most advanced of all heterotrophic organisms that cannot (as all autotrophs can) synthesize their own complex nutrients from simple inorganic inputs. Foodstuffs could be then seen as nothing but more or less complex assemblages of nutrients, and meat stands out among them for many reasons.

A small definitional detour is called for first because, as is often the case when dealing with seemingly straightforward subjects, everyday usage of the word "meat" does not coincide with biophysical realities. Meat, from a *sensu stricto* structural and functional point of view, refers only to the muscular tissue of animals, and the narrowest traditional definition would limit it to skeletal muscles of wild and domesticated mammals. Horowitz (2006) documents how even during the 1950s many American housewives did not consider chicken to be a meat and how the chicken industry was encouraged to run advertising campaigns that would confer on poultry a full meat

Should We Eat Meat?: Evolution and Consequences of Modern Carnivory,
First Edition. Vaclav Smil.
© 2013 John Wiley & Sons, Ltd. Published 2013 by John Wiley & Sons, Ltd.

status. There are also some national rules that make explicit definition. According to the Food Standards Code of Australia and New Zealand, meat is "the whole or part of the carcass of any buffalo, camel, cattle, deer, goat, hare, pig, poultry, rabbit or sheep, slaughtered other than in a wild state," a definition that pointedly excludes all wild species, including kangaroos whose meat is now readily available in Australia (Williams 2007).

In contrast, a common, *sensu lato*, usage extends the noun's coverage not only to muscles of all mammals and birds (much like the understanding of our pre-industrial ancestors for whom meat was everything from squirrels to bison and from thrushes to herons) but also to muscles of amphibians and reptiles (frogs, snakes, turtles) and to all other tissues that are often integrally or proximally associated with meat, above all to embedded or surrounding fat, sometimes also to skin and to internal organs (organ meats, innards, offal – *abats* in French, *frattaglie* in Italian, *Innereien* in German), most of which are not hard-working muscles. But even this liberal definition still leaves out all seafood although few skeletal muscles are as powerful and as efficient as those propelling fast cruising bluefin tunas that can (unlike all other ectothermic fish) raise their temperature above that of the surrounding water (Block 1994).

Nor is there any clear, universal divide between "red" and "white" meat. The distinction obviously owes to the amount of myoglobin in muscles (just 0.05% in chicken, up to 2% in beef), but because all mammalian meats have higher concentrations than poultry or fish, the USDA puts all large livestock meat into the red category. In contrast, the Australian definition of red meat refers to beef, veal, lamb, mutton and goat meat, but it excludes pork as well as all game meats, including buffalo whose meat is largely indistinguishable from beef. And then there is a common culinary usage that draws the line by age: veal, lamb and piglets are white; beef, mutton and pork are red, but so are duck and goose; and (to bring yet another color into the mix) in France, all game meat is labeled *viandes noires*. But lack of strict logic is common in classifying foodstuffs: tomato is, of course, a fruit that is always classified as a vegetable, to say nothing about counting tomato paste on pizzas as a vegetable.

Meat Eating and Health: Benefits and Concerns

In this introductory chapter, I will deal first with the functional and structural properties and the basic composition of muscles and other animal tissues before I turn to specific surveys of meat as a source of energy that comes (given the virtual absence of carbohydrates in muscles) only from two macronutrients, lipids and high-quality proteins. Most societies could always secure abundant, or at least adequate, amounts of carbohydrates from plants, but lipids, and even more so high-quality proteins, were relatively scarce in all traditional agricultures, as well as in the early stages of post-1500 modernization. That is why the role of animal protein in early human growth deserves particular attention.

Eating relatively large amounts of meat must have a variety of health and longevity consequences, but, as with all long-term effects of specific components of human diet, it is not easy to tease them out in an unequivocal manner from often inadequate and sometimes questionable epidemiological evidence. There is no doubt about the benefits of high-quality protein for young children in general and for their growing brains in particular, and there is also a high degree of consensus regarding the undesirability of consuming large amounts of fatty meat (although even here there are some intriguing caveats). More recently, a consensus has been emerging about the undesirability of frequent consumption of processed meat products ranging from bacon to wieners.

In contrast, solid generalizations regarding the contribution made by low to moderate meat consumption to the prevalence of the two leading causes of death in modern societies, that is, to cardiovascular and cancer mortality, are much more elusive – and hence it is difficult to say what might be the exact role of meat consumption in extending or reducing average human life expectancy. And, finally, when looking at links between meat and health, it is unavoidable to address the concerns about diseased meat, about meat-borne pathogens whose effects can range from mild individual discomfort to viral pandemics.

These risks have always been present in terms of bacterial contamination arising during the growth, killing of animals and post-slaughter treatment of carcasses and retail cuts, and several animal diseases with potential for epizootic outbreaks have always made their episodic appearance. But there have been two new developments during the past two decades: the emergence of contagious avian viruses with a strong potential for viral pandemics, and beef infected with a variant Creutzfeld–Jacob disease (vCJD) (human form of bovine spongiform encelopathy [BSE], commonly known as mad cow disease). Individual risks of the latter infection have always been minimal, but the avian influenza is a cause for legitimate worries as its future virulent manifestation can cause large global death toll.

Meat and its nutrients

Evolution has left us with no shortage of specialized organs to admire because of their intricate structures and amazing functions: brains and eyes are commonly cited as the pinnacles of evolution, but such rankings are meaningless as in living organisms only the synergy of all organs matters, and hence skins or intestines or bones or muscles are no less important. Muscles – the prime movers of heterotrophic locomotion that make all walking, running, jumping, swimming and flying possible – look macroscopically fairly simple, but viewing their structure sequentially upward from molecular level is a different matter (Aberle et al. 2001; Lawrie and Ledward 2006; Myhrvold et al. 2011).

Molecules of specialized proteins, actin and myosin, are organized in myofilaments that form sarcomeres whose contraction and relaxation generates all muscle motion. In turn, sarcomeres are grouped into myofibrils that are bundled into muscle fibers sheathed by a collagen matrix (endomysium); muscle fibers are bundled into fascicles that are contained within another collagen mesh (perimysium), and the entire muscle is covered by yet another collagen sheath (epimysium, or silverskin). The ends of these connective tissues merge into tendons that are attached to bones (but there are also some muscles that are not attached to skeleton). Tenderness of meat is determined by the size of fascicles (muscle grain) and by the strength and thickness of collagen sheaths. Coarser grain of more powerful muscles covered with stronger collagen results in less tender meat.

The division between light and dark meat reflects the muscle functions: rapidly twitching muscles, reserved for sudden, fast movements and brief exertion at maximum power, are lighter-colored, while the muscles for continuous but relatively low power exertions (breathing, standing, masticating) are composed of darker, slow-twitching fibers – they have more myoglobin, another specialized protein that moves oxygen from the blood to muscle cells. But there is no stark color difference in muscle color among those domesticated animals whose ancestors had large home territories or migrated over long distances: intermediate fibers of muscles in cattle or aquatic birds are all colored by myoglobin which accounts for 0.5% of muscle mass in cattle but for less than 0.1% in pigs.

Actin, myosin, collagen and myoglobin are all proteins (collagen is the most abundant protein in animal bodies), and hence muscles can be best thought of as intricate assemblies of wet proteins: on the average, living muscles contain about 75% water (extremes range from 65% to 80%), and their protein content is, at nearly 19%, the least variable major component; embedded lipids average about 3%, non-protein nitrogen (including nitrogen in adenosine triphosphate) is less than 2% and the small remainder are traces of carbohydrates (mainly glycogen) and inorganic matter (particularly iron and zinc). Because of their higher fat content, there is less water in animal carcasses (about 55% in beef and just over 40% in pork), but the protein content of their separable lean meat varies within a very narrow range, from 19% to 23%.

But most muscles also contain fat that is embedded in the sheathing collagen in order to supply long-acting aerobic fibers with a readily available and highly dense source of energy. This embedded fat also plays an essential role in meat's gustatory quality as it weakens collagen structures and makes meat more succulent, particularly once it degrades to gelatin during moist heat cooking once meat reaches 65°C. In contrast, no external application of fat can make a very lean meat as succulent as a more fatty cut, a reality that engendered a partial help through an ancient practice of larding lean cuts of meat. In some mammalian and avian species (particularly in such highly mobile wild animals as hares, deer or pheasants), there is only a small quantity of fat beyond the limited amount that is present in embedded stores, while in others there are substantial subcutaneous fat deposits as well as rich deposits surrounding internal organs.

Shares of separable lean and separable fat range widely among both beef and pork cuts. The extreme for beef are top round steak with almost 90% separable lean, just 8% of separable fat and about 2% of refuse when all fat is trimmed away, and short ribs with only about 40% of separable lean, 32% of separable fat and 27% of refuse (USDA 1992). Depending on taste preferences and health concerns, separable fat may be almost completely removed during butchering, preparation of retail cuts or final trimming before cooking, or it may be left in copious amounts on retail meat cuts and eaten as part of stews, roasts, barbecues or processed meats.

The heart is, of course, the only constantly working muscle in the human body, but among all other organ meats only tongue and gizzard are peculiar muscles (in the first instance, a complex network of muscles of great agility and omnidirectional mobility, in the second case an involuntary smooth muscle), while liver and sweetbreads (thymus) are enzyme-rich glands, tripe is a lining of ruminant stomach, and brain and kidneys are each *sui generis* organs. The composition of raw mammalian livers is very similar to that of skeletal muscles (about 70% moisture and 20–21% of protein), and tripe has about 19% of protein, but other innards are slightly to substantially less proteinaceous: kidneys and tongues have about 16% protein, hearts between 15% and 17%, sweetbreads 15% and brains only about 10% (and 80% moisture). Skin, contrary to common perception, has very high moisture content, and in some species (including pigs, chicken, ducks and geese), it is eaten as a part of broadly defined meat, either as crisply cooked part of meat in roasts or as a separate preparation.

Finally, all meat eaters also ingest some blood. Between 40% and 60% of all blood is lost by exsanguination and all but a small share of the rest is retained in viscera; as a result, the residual blood content amounts only to 2–9 mL/kg of muscle, and this minuscule rate does not appear to be affected by different ways of slaughter (Warriss 1984). When assuming mean blood content of 5 mL/kg, an annual consumption of 80 kg of boneless meat (recent US average) would imply annual intake of some 400 mL of residual blood. For comparison, the pastoral Maasai tribe in Kenya, who used to tap regularly the jugular veins of their cattle to drink blood or to collect it for mixing with milk, would draw at a time 4–5 L from a steer or a bull and half that volume from a cow or a heifer and consume several liters in a single month (Århem 1989). Maasai blood drinking has been in decline for decades, but in many societies blood is still consumed (albeit irregularly and in small amounts) in traditional dishes ranging from soups and stews to stir-fries and sausages. But a habit from the late 19th century is no longer with us: young Parisian women do not visit slaughterhouses to drink the blood of freshly killed animals in order to redden their cheeks (Gratzer 2005).

Although meat has been an important component of food energy supply during the long period of hominin evolution and a major contributor to energy intake in Paleolithic and Neolithic societies, its prime role was qualitative rather than quantitative: foods that are equally, or much more, energy-dense could be secured by gathering, but before animals were domesticated, and in societies that had limited access to aquatic foods, meat was the only source of the highest-quality protein. And while most wild animals have low, or even very low, deposits of fat, high energy density made animal lipids much sought-after, and only modern nutritional science discovered meat's value as an outstanding source of a key vitamin and of several essential minerals.

The physical and chemical properties of meat obviously determine its taste, ease of cooking, flexibility of preparation and hence the popularity of individual species or specific meat cuts. Nutritional composition is a different matter as the tissues and cuts that may rank low in terms of culinary preference may contain virtually identical shares of essential nutrients. Three kinds of preformed organic macromolecules present in plant and animal foodstuffs – carbohydrates, proteins and lipids – must be digested in relatively large quantities to serve as source of food energy, as well as sources of proteins and fatty acids that are indispensable for the growth and maintenance of human bodies. In modern diets, typical consumption rates of these macronutrients range from 10^1 g/day for proteins and lipids to 10^2 g/day for carbohydrates. In contrast, compounds and elements belonging to two distinct classes of micronutrients – vitamins and minerals – are ingested at low to very low rates, ranging from just a few grams per day for sodium and potassium to just a few micrograms per day for vitamin B11.

Meat contains virtually no carbohydrates, but it is an excellent source of high-quality proteins and fats. In those prehistoric societies that had no milking animals and no, or limited, access to aquatic species, meat was the only source of proteins needed for normal childhood and adolescent growth and adult body maintenance. The importance of meat in diets of hunters and gatherers encountered by the European expansion in the Americas, Africa, Asia and Australia has been abundantly described in the narratives of explorers and colonizers, and in the societies whose traditional way of life persisted into the 20th century, it was eventually studied and analyzed by modern ethnographers and anthropologists.

Some of these studies have included revealing quantitative analyses demonstrating the importance of domesticated pigs in New Guinea (Rappaport 1968), cooperative hunting among Tanzanian Hadza (Marlowe 2010) or dependence on collected and hunted wild animals among Ache of Paraguay (Clastres 1981). As I will show in some detail in Chapter 2, meat consumption declined to low or very low levels in all densely settled traditional agricultural societies, but during those millennia of low intakes, meat never lost its status of a highly desirable food. In the Western world of the 19th and the early 20th centuries, meat was valued both as a source of protein and fat, and its rising consumption was one of the major contributors to enhanced growth, increased adult weight and improved health of rapidly urbanizing populations.

Post-WW II affluence and new nutritional and health awareness changed the perspective: with the abundance of other high-quality protein sources (seafood, eggs, dairy products), meat lost its status of indispensable supplier of protein, and fatty meat (beef in particular) lost a considerable market share to lean pork and, above all, to chicken. The composition of meat consumption has changed, but in all modern societies, be they affluent Western countries or rapid modernizers of Asia, meat remains the single largest source of high-quality protein, followed by dairy products, fish and eggs (usually, but not necessarily, in that order). Meat also supplies significant shares of essential fatty acids and important micronutrients, above all iron – a mineral whose deficiency has been common in many populations, including women in affluent countries.

Few modern scientific advances have been as consequential as the discoveries of the importance of micronutrients to human health. Deficiencies of common minerals can impede normal human growth; low intakes of vitamins compromise essential metabolic functions ranging from gastrointestinal upsets to epithelial hemorrhaging. Balanced diets supplying adequate amounts of macronutrients in foods originating from a variety of plant and animal sources do supply sufficient quantities of micronutrients, but poor eating habits mean that even in the countries suffused with food and consuming excess of carbohydrates, fats and proteins, micronutrient deficiencies are common.

Iron deficiency is one of the most widespread as well as one of the most damaging problems as it affects as many as 1.6 billion people, or more than a fifth of all humanity (deBenoist et al. 2008), and, even more tragically, in low-income countries, it impairs brain development of roughly half of all children and is associated with every fifth maternal death (Micronutrient Initiative 2009).

Meat is one of the best sources of dietary iron because it supplies this essential mineral as heme iron that is easily absorbed in the upper small intestine and that also helps to absorb non-heme iron present in plant foods, and even modest meat consumption helps to prevent iron deficiency anemia (Bender 1992). Iron content in red meat is mostly between 1 and 2 mg/100 g; it is particularly high in mutton (more than 3 mg/100 g), and it is highest in organ meats (nearly as much as 10 mg/100 g in lamb liver and kidneys). Recommended daily intakes of iron are 8–11 mg/day for children and adolescents, 8 mg/day for adult men, 18 mg/day for pre-menopause women and 27 mg/day during pregnancy (Otten et al. 2006). This means that up to 25% of daily adult male requirements can be supplied by eating a single modest serving of red meat.

Zinc is the other metal present in relatively high concentrations. The element is a part of metalloenzymes (it is actually the most common catalytic metal ion present in cell cytoplasm), and as such it plays several essential roles in the synthesis of nucleic acids, protein and insulin.

Zinc-fingers (proteins containing the element in human genome) interact with DNA and mediate gene transcription. As with other metals, zinc from plant food interacts with phytate and becomes less bioavailable than zinc present in animal foodstuffs (as a result, vegetarians should ingest about 50% more than the standard recommendation). Zinc deficiencies include retarded growth, higher rates of infection, skin lesions and impaired wound healing and are a significant factor in poor world's morbidity. Largely vegetarian diets raise the molar phytate : zinc ratios to more than 20, or even 25, well above 15 (the threshold for predicting suboptimal zinc supply) and more than double the ratios of around 10 or lower that prevail in affluent countries (International Zinc Consultative Group 2004).

That is why this study group estimated that a quarter of all people in South and Southeast Asia and in Latin America are in a zinc deficiency risk category. But nutritional surveys have shown zinc intake below recommended intakes even among children and adults in affluent countries (Samman 2007). Those recommendations are 11 mg/day for adult males and 8 mg/day for females (Otten et al. 2006), while 100 g of red meat contains 4–4.5 mg of the metal (Williams 2007). Meat is also a good source of selenium and phosphorus. And given the concerns about excessive sodium consumption, it should be noted that meat is low in sodium and richer in potassium, with the ratio of the two elements ranging from 5 : 1 to 6 : 1. Meat contains no vitamin C, very low levels of vitamins A and D and very little of thiamin, but it is rich in three vitamins of the B group, in B6, B12 (particularly in organ meats) and niacin.

B6 (a group of six pyridoxine-related compounds) is a coenzyme essential for amino acid and glycogen metabolism, and its deficiency causes seborrheic dermatitis and microcytic anemia. Its daily requirements are 1.2–1.5 mg for adults; meat contains between 0.5 and 0.8 mg/100 g, but the main dietary sources of B6 in the Western diet are fortified cereals for females (Otten et al. 2006). B12 (cobalamin) is another essential coenzyme that is stored in large amounts in the liver; on a daily basis, it is needed only in minuscule quantities (adult intakes should be just 2.4 μg/day), and its deficiency, caused by interference with its complicated process of absorption, can eventually lead to megaloblastic anemia and neuropathy (Truswell 2007). Meat contains roughly as much B12 (1–2 μg/100 g) as cheeses or eggs, but its concentrations in livers and kidneys are an order of magnitude higher.

Meat as a source of food energy

When food supply is nutritionally balanced and adequate to meet growth, maintenance and activity needs, protein is only a marginal source of energy.

In such circumstances, (i.e., for the majority of today's populations), most of the needed food energy is always derived from carbohydrates and fats, and only a small, and globally fairly uniform, portion comes from proteins: among major economies, that share ranges only narrowly between 10% for India and 13% for France. In contrast, the other two shares range much more widely, between just 45% (France, Spain) and more than 75% (some sub-Saharan countries) for carbohydrates and between 10% (Ethiopia) and just over 40% (France) for fats.

As an energy source, protein has gross energy density about 35% higher than that of carbohydrates (23 kJ/g compared to 17.3 kJ/g), and those rates are less than half the value for energy-dense lipids (39 kJ/g). Actually metabolizable energy of carbohydrates is only marginally lower (16.7 kJ/g), but the adjustment for metabolizable energy – generally known as Atwater factor correcting for losses during digestion, absorption and urinary excretion (Atwater and Woods 1896) – is relatively higher for lipids (down to 37.3 kJ/g) and it is highest for proteins (down to 16.7 kJ/g). Only when the supply of carbohydrates and fats becomes inadequate, a rising share of food energy is drawn from protein metabolism, a condition known as protein-energy malnutrition that is still fairly common in many low-income countries. Global data about the extent of undernutrition and malnutrition are derived from statistical probability assessments, and hence no accurate figures are available, but the best published estimates put the extent of undernutrition at about 925 million people, or about 13% of the global population in 2010 (FAO 2010a).

Lean meat is much less energy-dense than the most common staple plant foodstuffs, cereals and leguminous grains, with energy densities mostly between 14 and 15 kJ/g of dry weight, and its closest plant counterpart in terms of energy content are sweet potatoes with 4.6–4.8 kJ/g. And while wild animals are generally leaner than their domesticated progeny, lean meat of all species has very similar energy density. Both lean beef and lean pork contain about 4.8 kJ/g, while venison and chicken meat average about 4.4 kJ/g compared to more than 5 kJ/g for more fatty cuts of red meats or for such large wild herbivores with substantial subcutaneous fat deposits as eland or moose (Eaton 1992). And while camel humps are significant fat depots, camel meat is also lean, with no inter- or intramuscular fat (Hertrampf 2004).

Lean meat is much more energy-dense than common leafy vegetables (cabbage and spinach at about 1 kJ/g) but comparable to sweet potatoes (4.7 kJ/g), and it is much less dense than all oil seeds, be they those with exceptionally high protein content (soybeans at 16.8 kJ/g) or oil content (sunflower seeds at more than 23 kJ/g). Another way to do this comparison (assuming, simplistically, that only energy matters) is by listing approximate mass of foods needed to satisfy daily food requirements of an active adult (12.5 MJ/day): it would be about 10 kg of leaves, more than 3 kg of fruits, 3 kg of tubers or white fish, about 2.5 kg of lean meat but less than 700 g of fatty pork.

Preferences have shifted: for most people, the gustatory appeal of meat, and in traditional societies (where fatty cuts were almost always seen as highly desirable) also the largest share of food energy, was associated with fat. Large domesticated mammals have three kinds of fat: substantial subcutaneous reserves serving as energy stores, deposits surrounding and cushioning such vital organs as heart and kidneys, and, as already explained, relatively small deposits of intramuscular fat. The first two categories of fat can be easily separated from skin, meat and bones and processed to yield edible products (particularly pig's back and belly fat turned into bacon or rendered into lard), added to such meat products as sausages or used as industrial ingredients (beef tallow).

Pork fat has less than 8% of moisture and a few percent of protein and energy density of about 34 kJ/g, while beef fat is even purer and has less water and less protein and energy density of nearly 36 kJ/g. Nearly 30% of beef carcass and almost 50% of pork carcass is fat, while separable lean beef averages less than 2% of fat, and lean pork, at just over 1%, is similar to venison; only lean veal and chicken have less than 1% of meat fat. Besides obviously raising a meat's energy density, the presence of fat affects flavor and juiciness of meat and helps to create satiety, a feeling that prolongs the intervals between meals and leads to a common preference for such foods (Rogers and Blundell 1990). But, contrary to a common perception, the presence of fat is not the primary determinant of tenderness (the size of muscle fibers is).

Trend toward leaner cuts, driven above all by health concerns, has been unmistakable in all affluent societies, and meat producers have responded in three principal ways: by selective breeding and adjusted feeding practices producing leaner carcasses, by meat classification and marketing that favor leaner cuts, and by improved butchery techniques. As a result, the share of separable fat in retail meats has been declining, and Williams (2007) reported that for Australia's almost exclusively grass-fed animals, all trimmed beef cuts now have less than 5% of total fat, trimmed lamb cuts average less than 10% of fat and the only meat with more than 10% fat is regular mince. Effects of trimming and cooking on the fat content of actually eaten meat can be profound: for example, Gerber et al. (2009) found that cooking beef, veal and pork reduced the absolute fat content by 18–44%, and trimming of edible fat cut away an additional 24–60%.

But some societies and some consumers still prefer those beef cuts with high, even very high, proportion of intramuscular fat. In Japan, a country of restrained meat consumption, by far the most expensive domestically produced meat is beef that comes from the crosses of traditional black-coated, small-stature breeds of *wagyū* (Japanese cattle used for centuries as draft animals) and European animals. Kōbe beef from Hyōgo prefecture is produced by feeding penned heifers for up to three years, regularly massaging them and giving them beer to drink during the summer, and it is so highly marbled that it contains 20–25% of fat compared to 6–8% for the USDA prime beef.

The most expensive cuts look more white than red and are sold for more than $500/kg. And the even more expensive Matsuzaka beef (from Mie prefecture) has an extremely high fat/meat ratio; the animals are also fed beer in summer but are also massaged with *shōchū* (Japanese liquor). But true *wagyū* beef remains a niche market category, and in affluent countries, the common perception of animal fats has undergone a negative transformation during the past two generations: fatty meat used to be seen as desirable, deliciously filling, too often beyond the means of low-income households. Processed meat products – from French *patés* to numerous varieties of Italian *salami* – are held by many in high culinary esteem and contain a very high proportion of fat, often in excess of 40%.

Dietary recommendations by FAO/WHO experts set the acceptable range of total fat intake at 20–35% of all food energy, with minima at 15% in order to ensure not only adequate total energy intake but also the supply of essential fatty acids and bioavailability of fat-soluble vitamins A, D, E and K (FAO 2010b). Dietary guidelines for Americans recommend keeping the total fat intake between 20% and 35% of food energy while maximizing the share of polyunsaturated and monounsaturated fatty acids and consuming less than 10% of daily energy in the form of saturated fats (USDA 2010a). Recent shares of fats in average food energy supply of affluent nations have ranged from 25% to 27% in Japan (the lowest share among all affluent countries and the same as in China) to as much as 41% in France and 38% in the US (FAO 2012). Oils (corn, soybean, rapeseed, olive, etc.) are the largest source of plant fats; in Western countries, a large share of lipids came from dairy products, and the recent contribution of meat fats was less than 10% in the West (about 9% in France, 7% in the US).

Biochemically, there are three kinds of meat fats, triglycerides (3 moles of fatty acids joined to glycerol), phospholipids and cholesterol, an essential dietary ingredient involved in hormonal function and indispensable for cell wall integrity. Fatty acids are conventionally grouped into saturated (SFA), monounsaturated (MUFA) and polyunsaturated (PUFA) categories, but this broad chemical classification ignores the distinct biochemical properties of individual acids. Modern dietary research has convincingly established that it is desirable to limit the intake of SFA and replace it by MUFA (in practice, this means by consuming more oleic or erucic acid, that is, olive and rapeseed oil) and PUFA (present in fish). Dietary recommendations for adults are to limit SFA to 10% of all energy intakes while consuming at least 6% and up to 11% of all energy as PUFA, especially as n-6 PUFA and n-3 PUFA (FAO 2010b).

A common misconception has been to see meat as a dangerously high source of SFA, while their amount, even in red meat, is actually lower, per edible portion, than the combined quantities of MUFA and PUFA. According to the USDA standard reference data base, the share of SFA is 45% in beef tenderloin, 44% in top round steak, 38% in composite retail cuts of trimmed pork, 35% in skinless chicken breast and 32% in skinless chicken thigh (USDA 2011). For comparison, SFA make up 30% in albacore tuna, 22% in salmon and 14% in olive oil. Given the variety of cuts, animal breeds and feeding regimens, it is hardly surprising that fatty acid analyses done in Brazil or Australia are not identical with USDA's averages. For example, Williams (2007) found that SFA share in Australia's beef was 41%, while the results obtained by de Almeida et al. (2006) for Brazilian beef emphasize the often large difference among specific cuts, with bottom round steak having 45% SFA and top round steak 52% SFA, but the bottom round containing three times as much SFA in absolute terms.

Comparisons of the just-cited US shares with Australian and Brazilian values for grass-fed animals do not indicate that grazing lowers SFA content. Just the opposite was found by Rule et al. (2002) who examined fatty acids in free-ranging and feedlot-fed bison and beef, but they also found that the grass-fed animals have higher shares of PUFA. Similarly, Leheska et al. (2008), who analyzed US ground beef and strip steaks by using samples from 15 grazing operations in 13 states and comparing them with feedlot beef, found that grass-fed ground beef had significantly more SFA (55% compared to 47%) and significantly less of more healthy MUFA (42% compared to 50%) than did the feedlot beef. For strip steaks, the difference was smaller (52% SFA for grass-fed and 48% SFA for feedlot beef), while the concentrations of PUFA did not differ between the two kinds of meat. In contrast, several studies found that increased grass intake led to decrease of SFA concentrations (French et al. 2000). Ruminant meats (and dairy products) also contain conjugated linoleic acid, a naturally occurring *trans* fat shown by *in vitro* and experimental animal studies to have potential health benefits due to its cancer-inhibiting effect at several sites, above all in the mammary gland (Bhattacharya et al. 2006).

While many fatty acids are easily substituted by other structurally similar compounds, there are two essential fatty acids that must be present in all healthy diets: linoleic acid (a primary ω-6 PUFA) and α-linolenic acid (an ω-3 PUFA) must be digested completely preformed in order to become precursors of prostaglandins (that act as regulators of gastric function, smooth-muscle activity and hormonal release) and parts of cell membranes. Inadequate supply of these acids is manifested in scaly rash, dermatitis, neural abnormalities and reduced growth. US guidelines for linoleic acid recommend average per capita intakes of 17 g/day for adult men and 12 g/day for adult women, while the rates for α-linolenic acid are set at, respectively, 1.6 and 1.1 g/day. Red meat, with between 0.3 (beef) and 0.4 g (mutton) of ω-6 fatty acids per 100 g of edible portion, is as good or a better source of these nutrients as oily fish, but ω-3 acids are far more abundant in oily fish (more than 2 g/100 g compared to just 0.1–0.2 g/100 g).

Cholesterol is a part of cell membranes, and hence its content in the muscles of all mammals is fairly similar: contrary to common perception, beef or pork are not exceptionally high in cholesterol. The actual averages are mostly between 50 and 75 mg/100 g for well-trimmed beef and pork and 60–80 mg for chicken (de Almeida et al. 2006; Williams 2007). Cholesterol concentrations in red meat are higher than for some wild buffaloes and pheasants (both less than 50 mg) but lower than many published values for antelopes, deer and caribou that range, depending on the species, between 80 and 110 mg (Medeiros et al. 2001; USDA 2011). According to the US dietary guidelines, daily cholesterol intake should be less than 300 mg/capita (USDA 2010a). If meat were the only source of dietary cholesterol, this would translate to a maximum of 400 g of beef a day.

Meat's share of overall food energy supply was inevitably low in early hominins devoid of suitable hunting and butchering tools. It reached its evolutionary peak in Pleistocene hunting societies skilled at killing many species of (now mostly extinct) megafauna, was much lower in early sedentary societies combining foraging and cultivation of crops, and reached its lowest levels in all traditional agricultural societies practicing intensive cropping (and virtually absent in some of them). In Western societies, it began to rise with the 19th-century urbanization and industrialization, and in most of them it rose to averages unprecedented in history (albeit still lower than prehistoric means).

All of these supply and consumption shifts will receive detailed quantitative coverage (based on the best available evidence) in Chapters 2 and 3, while in this section I will briefly survey energy densities of all common species and many favorite cuts of meat. Here, I will just note the recent shares of meat in the total retail-level supply of food energy in the world's largest economies: for affluent nations, they range, much like the shares for protein, only narrowly from 11% as the EU mean to 12% in the US and 14% in Spain; remarkably, China's meat share in food energy supply is now as high as in Spain, but the share remains low in Japan (a bit over 6% because of the country's traditional preference for seafood), and it is minuscule (about 0.6%) in India.

High-quality protein and human growth

Meat's importance in human diets is primarily due to the supply of high-quality protein, secondarily to the provision of essential fatty acids and micronutrients and finally as a source of food energy. Dietary proteins are indispensable for all heterotrophic growth, including the maintenance and replacement of tissues. Nine amino acids are not synthesized by humans and must be ingested fully preformed: these essential amino acids (histidine, isoleucine, leucine, lysine, methionine, phenylalanine, threonine, tryptophan and valine) contain on the average about 16% of nitrogen and are irreplaceable precursors of all structural and functional proteins that form skeletal and other muscles, internal organs and bones as well as all complex, metabolically active compounds (enzymes, hormones, neuro-transmitters and antibodies).

Dietary proteins are also needed in order to make up for small but constant nitrogen losses due to the shedding of skin particles, cutting of hair and nails, and excretions in urine (Pellett 1990). Obligatory nitrogen excretions dominate with between roughly 40 and 70 mg/kg in adult-hood (mean of just over 50 mg/kg); other losses amount to less than 10 mg/kg. Recommended daily intakes are quantified in terms of reference (or ideal) protein that combines the presence of adequate amounts of all essential amino acids with easy digestibility. Chicken egg or cow milk protein have been the two most common whole food choices, but any animal protein has the same high ranking, and so any meat or any fish could be used as a source of ideal protein as well.

Obviously, inferior proteins that contain suboptimal amounts of one or more essential amino acids as well as those that are difficult to digest cannot support the same rate of growth. In vegetarian diets, protein quality is usually most affected by lysine deficiency (this essential amino acid is present in relatively low amounts in all cereal grains) or by shortages of sulfur-containing methionine and cystine (whose levels are relatively low in all leguminous grains). In everyday diets, complete (ideal) proteins (with more than adequate shares of all essential amino acids) are available only in foods of animal origin (meats, fish, eggs, dairy products) as well as in mushrooms, but all plant foods have incomplete proteins (with one or more amino acids relatively deficient).

This means that the scores for protein quality of common vegetarian diets based on grain or tuber staples will be only around 70 and even as low as 60 compared to 100 for a reference protein, and that in order to receive adequate intake of all essential amino acids, infants will have to consume 40–70% more of protein in their meatless diets than they would have to eat in a mixed diet containing some dairy products, meat or fish. The second key variable is actual digestibility of proteins. The ratio for egg and dairy protein is, respectively, 97% and 95%, and at 94% the rates for meat and fish are nearly as high (FAO/WHO 1993). In comparison, digestibility for whole wheat, corn and oatmeal are 85–86% and for beans less than 80%, and the rates for legume-rich mixed diets are as low as 77–78% in India and Brazil.

Since the early 1990s, the preferred method for evaluating the protein quality of foods has been to use protein digestibility-corrected amino acid scores (PDCAAS). These scores take into account age-related scoring patterns of amino acid requirements (different for children and adults) and adjust them for digestibility (FAO/WHO 1993). Reference PDCAAS for casein and egg white is 1.00, beef scores 0.92, chickpeas around 0.7 but lentils, as well as whole wheat, only 0.52 (Sarwar et al. 1989). Mixed diets including animal foods, where all amino acids are always present in more than adequate amounts, will get fairly high scores; for example, combining whole-wheat products and beef raises PDCAAS to 0.85, and for a typical Western diet based on refined wheat flour, the ratio will be above 0.9.

After decades of studies, we have a fairly good understanding of protein and amino requirements in human nutrition. The FAO and WHO specify the "safe level of intake" (i.e., the minimum needed to maintain protein balance) at 0.83 g/kg/day for adults of both sexes (FAO/WHO 2007). A woman weighing 50 kg will thus need 42 g of protein a day, and a 75-kg man will require 62 g/day. For infants, these recommendations fall from 1.31 g/kg/day at six months of age (or about 10 g/day) to 1.14 g at one year and to 0.9 g/kg/day at the age of ten years, translating to an average daily intake of about 25 g/capita. Additional protein is needed during pregnancy and lactation. For comparison, the US dietary reference guidelines estimate the average daily protein requirement at 0.66 g/kg for adults and set the recommended daily allowance at 0.80 g/kg/day for both sexes (FNB 2005), while the reference values for Australia and New Zealand are 0.84 g/kg for adult men and 0.75 g/kg for adult women (NHMRC 2006).

All of these values refer to ingestion of proteins with a digestibility-corrected amino acid score value of 1.0, which means that appropriately higher amounts will be needed with diets whose proteins have inferior PDCAAS. These conservatively specific requirements are well below the values of the amount of protein that is actually available in average per capita food supply not only of all affluent countries but also many populous modernizing nations. During the first decade of the 21st century, average daily protein supply was in excess of 100 g/capita both in the US and France; it was around 90 g/capita in Japan as well as in China with Brazil not far behind (FAO 2012). Meat supplied roughly a third of all protein (and more than half of all high-quality protein from animal foodstuffs) in the US as well as in Brazil, more than a quarter in France and a fifth in China.

Europe is the only continent where average per capita consumption of meat protein has become saturated at about 25 g of protein a day by the end of the 20th century; all other continents have seen steady increases to maxima of nearly 40 g/day in Australia and 30 g/day in the Americas, rates that are an order of magnitude higher than in the world's least developed countries (FAO 2012). Among the world's most populous nations, only the overwhelmingly vegetarian India and Bangladesh stand out with less than 60 g/day of total protein supply, with meat contributing just 2% in India and less than 3% in Bangladesh. Given the combination of inevitable nutrient losses along the food chain, of low amino acid scoring, of poor digestibility of the subcontinent's legume-dominated plant proteins and of unequal access to food, it is obvious that a daily protein supply of less than 60 g/day is, at best (among better-off social strata and in some regions), barely adequate, while hundreds of millions of people remain undernourished: not surprisingly, in absolute terms India has the largest number of malnourished people (FAO 2010a).

One of the more intriguing recent studies has been an investigation of meat eating and cognition in China's Guangzhou province (Heys et al. 2010). When compared to no meat consumption in childhood or to eating meat just once a year, daily meat eating was positively associated with both immediate and delayed recall score in a study of more than 20,000 Chinese men and women aged 50 years and over. Studies in Nepal and Kenya demonstrated a similarly beneficial effect on motor milestone acquisition and on growth and cognitive function among children (Neumann et al. 2003; Siegel et al. 2005). If further confirmed, these findings would be of great importance, given that some 60% of the world's people with dementia now live in modernizing countries where meat intakes are often very low.

On the other end of the global nutritional spectrum are societies with a surfeit of food and high incidence of obesity: there, too, adequate protein intakes can be beneficial. Because dietary protein is more satiating than carbohydrate or fat, diets with more protein are more likely to reduce food intake and result in greater weight loss than high-carbohydrate diets; eating lean meat can thus help to reduce the rates of obesity and type 2 diabetes (Noakes et al. 2007).

Carnivory and civilizational diseases

The pattern of morbidity and mortality had changed during the latter part of the 20th century, as inoculation reduced and eventually eliminated the incidence of once common infectious diseases, as antibiotics became widely available to treat life-threatening bacterial infections, and as better sanitation and stricter preventive legislation cut the risk of food poisoning and dangerous exposure to environmental pollutants. As a result, the Western world and, with only a slight delay, the urbanized areas of lower-income countries have experienced the rising frequency of so-called civilizational diseases, illnesses whose genesis is associated with lifestyle – including diet, stress and lack of physical activity – and whose incidence increases with greater longevity.

The list of these diseases ranges from asthma to osteoporosis, but the two largest categories (and also the two leading causes of death in modern societies) are cardiovascular diseases (CVD, including cardiac and vascular mortality, or its major subcategory, coronary heart disease, CHD) and cancers. Diet, in general, and meat consumption, in particular, have been singled out as major contributors to the genesis of CVD (among recent studies, see Williamson et al. 2005; Kontogianni et al. 2008). But critical reviews of these statistical associations do not reveal any convincing causation at low to moderate levels of meat consumption, and the overall statistical association weakens considerably once the fat is separated from meat (Li et al. 2005; Givens 2010; McAfee et al. 2010).

The now classic link between the consumption of meat (fatty meat to be precise) and higher incidence of CVD mortality was established by the Seven Countries Study that focused on the links between CHD and lifestyle factors, particularly dietary fat intake (Keys 1980; Alonso et al. 2009). The study, whose baseline surveys were conducted between 1958 and 1964, included 16 cohorts of men between 40 and 59 years of age in seven countries, and its most widely reported outcome was a strong positive correlation between the average intake of saturated fats (coming from meat as well as from separated animal fats and eggs and from hydrogenated plant oils) and CHD mortality.

The difference between low- and high-fat diets was particularly well demonstrated by contrasting American and northern European cohorts with two Japanese cohorts, the first one from a farming village of Tanushimaru, the second one from a predominantly fishing village of Ushibuka. Men in the Japanese villages had age-standardized 25-year CHD mortality between just 30 and 36/1,000 compared to rates in excess of 100/1,000 for their US and northern European counterparts. Only a cohort from Crete had a lower mortality rate at 25/1,000, while the mean for Mediterranean diets was 40–90/1,000 (Menotti et al. 1999). The most significant predictors of high CHD mortality were the average intakes of butter, lard, margarine and meat, while higher consumption of legumes, oils and alcohol had the strongest negative correlations. And the continuing contrast between the Japanese and Western diet appears to confirm the fat–disease link, as average intake of fats is more than a quarter of all food energy in Japan compared to a third or more of the total in the West, and the Japanese CVD mortality is significantly lower.

But what was once a widely accepted epidemiological dogma is now anything but that. The obvious qualification that should have always been made was that the link was between fats and CVD, not between lean meat and CVD, and that the intakes of solid saturated fats (butter, lard, margarine) had the highest correlations with disease frequency. But more important facts for deconstructing even that simple fat–CVD link became soon available. Curiously, the Seven Countries Study did not include any men from France, and hence its result had entirely missed what came to be known as the French paradox, namely, the coexistence of low CHD mortality with high intakes of saturated fat and dietary cholesterol (Renaud and de Lorgeril 1992).

The prevalence of this paradox was soon found in other parts of Mediterranean Europe (Masia et al. 1999), and its initial explanation attributed the effect primarily to frequent drinking of red wine. The latest explanations also take into account composition of the entire diet with high fruit and vegetable intakes and regular physical activity (Ferrières 2004). And Spain's rapid dietary transition after the collapse of Franco's regime and accession to the EU provided strong support for the Spanish paradox: even as the country's per capita meat consumption rose to be the highest in the EU (accompanied by increasing dairy intakes), its CVD mortality decreased after 1976 (Serramajem et al. 1995).

Finally, another study documented what might be labeled the Japanese paradox: average per capita intakes of fats, meat and dairy products were increasing during the 1960s, 1970s and 1980s and led to higher mean blood cholesterol levels, higher average body mass, higher mean blood pressure and higher incidence of overweight – but the CHD mortality remained the same (Toshima et al. 1994). The validity of this paradox has been extended to the 1990s by showing that despite similar cholesterol levels and similar blood pressures among Japanese and American men, the Americans had CVD mortality about twice as high as the Japanese (Sekikawa et al. 2003).

By now, we have sufficient evidence to make several important conclusions about the links between meat and CVD. Modern lean red meat trimmed of visible fat has low content of saturated intramuscular fat – even beef muscle has less than 5% fat, and marbling fat concentrations as low as 20–50 g/kg are possible (Scollan 2003) – and low cholesterol content, and its low to moderate consumption does not raise total blood cholesterol and LDL cholesterol levels (Li et al. 2005; McAfee et al. 2010). People eating lean beef, pork and chicken in addition to typical Western diet will get much more fat (total and saturated) from fast and snack foods, oils, spreads and baked goods. Moderation makes a critical difference even when assessing the link between all red meat (lean, fatty and processed) and CVD.

Many studies that looked at red and processed meat consumption and CHD have been poorly designed, and this limits their confident interpretation. While most of them controlled for the main confounding variables (age, body mass index, alcohol, smoking, physical activity, etc.), they do not give absolute figures for the amounts of meat associated with higher CVD risks and use instead inconsistent "servings" or "portions" whose size is particularly difficult to standardize when the studies rely on dietary recall; another questionable approach is to calculate the degree of risk by contrasting the lowest and the highest quintiles of meat intake, a choice that excludes most of the people who consume moderate amounts.

The most comprehensive meta-analysis of these studies, based on 20 publications whose adequate design qualified for inclusion, found that red meat intake was not associated with CHD (relative risk of incident CHD when eating 100 g/day of red meat was 1.0), that processed meat intake had a strong correlation (relative risk of 1.42 when eating 50 g/day) and that total meat intake had an intermediate association, with relative risk of 1.27 when consuming 100 g/day of all meats (Micha et al. 2010). But as Bryan (2011) pointed out, such risk quantifications leave us confused: if a 42% higher relative risk that might arise from eating processed meat should lead to reduced meat consumption, what is then the lesson of the European EPIC-Oxford study that found vegetarians having increased colon cancer incidence (relative risk 1.39) when compared with meat eaters (Key et al. 2009)? That we should eat fewer vegetables?

The largest study included in the just-cited meta-analysis followed more than half a million Americans (aged 50–71 years at baseline) for 10 years, and it found that the risk of CVD was modestly elevated for both men and women only in the highest quintile of red and processed meat consumption in which the average intakes for men were, respectively, more than three times and nearly three times than in the second quintile (Sinha et al. 2009). Men in the highest quintile ate 68.1 g/1,000 kcal, that is – assuming about 2,300 kcal/day – about 55 kg of red meat a year compared to only about 8 kg/year in the lowest and 18 kg/year in the second lowest quintile. Links to stroke also remain uncertain: three studies found no association, while a Swedish study of women found that total red and processed meat consumption carried a significantly increased risk of cerebral infarction but not of total stroke or cerebral hemorrhage (Larsson et al. 2011), while for men there was no association with the fresh red meat and a positive link between processed meat consumption and stroke (Larsson et al. 2011a).

Most epidemiological studies of links between meat consumption and cancer – possible biochemical explanations of such causation are reviewed in Ferguson (2010) – have focused at red and processed meat and colon cancer: links with gastric cancer, a disease rare in rich countries, is not convincing, and those with breast and prostate cancers did not show up in large epidemiological studies (Corpet 2011). But a meta-analysis of ten studies concerning the association between breast cancer and red meat consumption in premenopausal women suggested the summary relative risk at 1.24, with 1.57 for case-control studies and 1.11 for cohort studies (Taylor et al. 2009). A number of suggested pathways that link red meat consumption and breast cancer involve hormonal action, and this would indicate a possible role of meat eating in the increasing incidence of hormone receptor-positive breast cancers documented in the US population since the early 1990s.

But a link between red meat intake and the colorectal cancer is seen as much more convincing, and a recent summary of this evidence led the World Cancer Research Fund (WCRF) to recommend limited consumption of red meat and avoidance of processed meat (WCRF and AICR 2009). Corpet's (2011) experimental studies with rats concluded that this is due to a true causative association and not due to confounding factors. But, as with the CVD link, two meta-analyses of more than 40 studies present more qualified conclusions (Norat et al. 2002; Larsson and Wolk 2006). Most notably, total meat consumption (red, white, processed) is not linked to colorectal cancer risk. When compared with consumers in the lowest quintile, the relative risk is significantly higher for people consuming the largest quantity of red meat, and similar level of risk (about

1.3) applies to eating processed meat. Again, the quantities make all the difference: WCRF and the American Institute for Cancer Research (2009) recommend the limit of 500 g/week of fresh red meat, or an average annual per capita consumption of up to about 25 kg.

And the latest prospective evaluation – using the Health Professionals Follow-up Study for men (1986–2008) and Nurses' Health Study for women (1980–2008) – looked at red meat consumption and mortality due to both CVD and cancer, and it has only confirmed the conclusions reached since 1990 (Pan et al. 2012). After correcting for major lifestyle and dietary risk factors, it found, once again, a linear dose response with the risk ratio of total mortality averaging 1.13 for one serving a day increase for fresh meat (standard serving size being 85 g, or a cumulative intake of 31 kg/year) and 1.2 for processed red meat, with specific risk ratios at 1.18–1.24 for CVD and 1.10–1.16 for cancer mortality. The author also estimated that substituting one serving of meat per day by other foods would lower mortality risk by 7–19%, and that the overall mortality could be reduced by about 9% for men and almost 8% for women if everybody consumed no more than half a serving of meat a day.

Despite many specific uncertainties, the cumulative epidemiological evidence is thus fairly conclusive, both in cases of CVD and cancer links to meat consumption: moderate meat intakes are the optimal choice to counsel, particularly when considering the benefits of ingesting complete proteins and easily absorbable micronutrients. Given a diverse diet, moderate food energy intake may be a much more important determinant of health and longevity than a particular dietary composition; this benefit is due to a positive effect of caloric (or dietary) restriction, a matter to which I will return in Chapter 5.

Diseased meat

Concerns about the safety of meat for human consumption have always included the risk of contamination due to improper slaughtering, storage or processing procedures and the presence of natural pathogens. Despite the advances in public hygiene and stricter rules for production and food treatment, the risk from natural pathogens remains common in the 21st century (Sofos 2008). Trichinellosis is the most recurrent problem among pigs and foot-and-mouth disease among cattle as well as pigs (Pozio 2007). Human trichinellosis is acquired by eating raw or inadequately cooked pork that harbors larvae of *Trichinella spiralis*, a small tissue-dwelling nematode that lives in domestic and wild animals in all inhabited continents and whose adult worms colonize human duodenum and jejunum. Worldwide incidence of human trichinellosis is underreported, but annually there are at least 10,000 cases with low (about 0.2%) rate of mortality.

Not surprisingly, trichinellosis has been relatively frequent in the world's largest pork-eating nation, with the outbreaks during 2000–2003 reaching nearly 1,000 cases and causing 11 deaths (Cui et al. 2011), while in the US less than 50 cases have been reported recently per year. Brucellosis, caused by different species of fever-inducing genus *Brucella*, has been similarly rare. These bacteria can infect all domestic animals but are not usually found in meat, and their transmission to humans has been greatly reduced with pasteurization of milk (Franco et al. 2007). The annual incidence in the US has been recently less than 100 cases.

Foot-and-mouth disease (also known as hoof-and-mouth disease) is due to a highly contagious virus of the Picornaviridae family (*Aphtae epizooticae*). Infected animals have high fever, foamy salivation and blistered feet. Fortunately, human infections are very rare, but the economic impact of regional or national epizootics is considerable. By far the most extensive recent foot-and-mouth disease epizootic began in England in February 2001, and the eventual infection of more than 2,000 animals led to mass slaughter of about seven million sheep and cattle and huge pyre burning of their carcasses, with estimated losses of about $16 billion (Ferguson et al. 2001). More recent, and less severe, outbreaks took place in parts of China in 2005, yet again in the UK in 2007 (confined to a small area in Surrey), but another nationwide infestation, that began in November 2010 in South Korea, led to a slaughter of some three million pigs and more than 100,000 cattle.

Infestations of meat by commonly occurring bacteria pose the most frequent risk and create the highest public health concern. By far the most common pathogenic bacteria ingested with meat belong to a ubiquitous species *Escherichia coli*, whose hundreds of strains reside without any ill effects in human and animal intestines. For example, among nearly 12,000 meat samples collected from four US states between 2002 and 2008, more than 80% of chicken and turkeys, nearly 70% of beef and more than 40% of pork were contaminated (Zhao et al. 2012). An overwhelming majority of these bacteria cause no problems and perish during cooking, but in 1982 a virulent strain O157:H7 of Shiga-toxin producing *Escherichia coli* (STEC) was first identified in contaminated and undercooked hamburger meat in the US.

Its ingestion may cause only a temporary discomfort in healthy adults, but it can result in severe illness (often marked by bloody diarrhea) among healthy people and a rapid death in children due to hemolytic uremic syndrome that leads to acute kidney failure. The Centers for Disease Control and Prevention estimates that in the US there are annually more than 73,000 STEC infections resulting in more than 2,000 hospitalizations and 60 deaths, and Frenzen et al. (2005) put the annual cost of STEC at nearly half a billion dollars.

But the foodborne bacterial pathogen that causes most illness, hospitalizations and deaths is caused by nontyphoidal *Salmonella*: in 2011, there were more than one million cases of illness, nearly 20,000 hospitalizations and close to 400 deaths (CDCP 2012). *Salmonella enterocolitis* infects the lining of the small intestine, and it is the most common cause of food poisoning in affluent countries. Abdominal and muscle pain, chills, fever, diarrhea and vomiting usually go away after two to five days, but dehydration is a dangerous risk in small children and infants (Pegues and Miller 2009). Improper handling and storage of poultry and eating of undercooked chicken and turkey are the most common sources of bacteria. Species of *Campylobacter* are responsible for more than 800,000 illnesses and some 70 deaths every year.

Salmonella occurrence is common not only on raw chicken in low-income countries but also in Europe and North America, with recent studies finding the prevalence of 68% in Addis Ababa, 66% in Bangkok, 60% in Oporto (Portugal), 36% in Belgium and Spain, and within a range of 39–65% in six of China's provinces and in Beijing and Shanghai (Yang et al. 2011) – while the old US standard tolerated *Salmonella* presence in up to 23.5% of samples of carcass rinses. In 2010, the USDA issued a new performance standard that limits *Salmonella* contamination of raw chickens to 7.5% of samples tested, and it also set up, for the first time, standards for *Campylobacter* genus; the two measures were expected to prevent nearly 40,000 of *Campylobacter* infection and 26,000 cases of salmonellosis (USDA 2010b).

These statistics must be seen in a realistic risk perspective. More than 300 million Americans eat meat, and hence four meals a day (breakfast, lunch, dinner, snack) would imply 1.2 billion meals that could contain some meat. Even when assuming that all *Escherichia*, *Salmonella* and *Campylobacter* infections come from meat – a great exaggeration as *Escherichia* is often ingested in water, juices, vegetables, fruits and milk, and *Salmonella* poisoning often comes from milk, eggs and vegetables – the annual total of roughly 1.9 million cases would mean that the risk of getting ill would be when eating roughly one out every thousand meals.

One out of every 40,000 meals would be followed by hospitalization, and one out of 2.2 million meals would carry a risk of foodborne death – and those roughly 500 deaths should be compared with about 20,000 deaths caused in the US by the annual influenza and with nearly 100,000 deaths due to hospital-acquired bacterial infections (Peleg and Hooper 2010). Moreover, proper meat handling in kitchen and cooking to recommended temperature reduce these risks to negligible rates. That is why an entirely different disease-related risk is much more worrisome: antibiotic-resistant bacteria in meat have potentially life-threatening and economically costly consequences.

Experiments dating to the late 1940s discovered that antibiotics digested with feed boosted weight gain of broilers by at least 10%, and already in 1951 the US Food and Drug Administration (FDA) approved the use of two common compounds (penicillin and chlortetracycline) as commercial feed additives, with oxytetracycline following in 1953, and these compounds have been used as inexpensive growth enhancers by the entire livestock industry. By the end of the 20th century, US poultry producers were using more antibiotics than either pig or cattle growers, but the total use of antibiotics in American livestock has remained contested for decades. In 2001, the Union of Concerned Scientists claimed that the past published totals of use were drastic (almost 50%) underestimates and put the annual US consumption at 11,150t (UCS 2001). The best available recent summation by the FDA (2010) is about 13,000t in 2009, or about 80% of the country's total use of these compounds.

Overuse of antibiotics increases the risks of widespread occurrence of antibiotic-resistant bacteria, particularly of the ubiquitous *Escherichia* and *Salmonella*. Zhao et al. (2012) found that in nearly 12,000 meat samples collected between 2002 and 2008, half of *E. coli* bacteria were resistant to tetracycline, more than a third to streptomycin and nearly a quartet to ampicillin, all commonly used to treat people. These decades-old concerns are being addressed by more restrictive rules. In early 2012, the FDA announced its prohibition of any prophylactic use of cephalosporins (antibiotics commonly used to treat human infections, including a still common pneumonia) in livestock and limited the use in farm animals to only two cephalosporin compounds (Gilbert 2012). Perhaps the most worrisome is the recent finding that nearly half of all meat sold in US supermarkets is contaminated with *Staphylococcus aureus*, whose specific genotypes in different meats point to its origin in the animals rather than in the human handlers (Waters et al. 2011).

This species is now infamous for its high degree of antibiotic resistance (96% of samples in studied meats were resistant, more than half of them to at least three different drugs), and its methicillin-resistant strains (MRSA) pose a greater danger to hospitalized patients than their illnesses or operations because that drug has been, in many cases, the last effective treatment. Until recently, MRSA findings in animals were limited to dairy cattle with mastitis, but since 2005 a bacterial clone (CC398, whose origin remains unknown) has been colonizing pigs, calves, dairy cows and broilers (Vanderhaeghen et al. 2010). Obviously, a possibility of this clone, and other virulent microbes, spreading to humans is a major concern. Animal-to-human spread of antibiotic resistance can take place by direct contact with animals as well as through the food chain, and Marshall and Levy (2011) summarize well-documented cases of such transmissions. Modeling suggests that the appearance of antibiotic-resistant commensal bacteria in humans has the greatest impact in the earliest stage of emerging resistance (Smith et al. 2002).

As if these risks were not enough, our dubious commercial choices have created an entirely new disease risk by converting cattle, that paragon of slowly chewing herbivores, into cannibalistic carnivores eating the rendered bodies of their deceased conspecifics. This unnecessary but common practice has been responsible for the genesis of BSE, commonly known as mad cow disease. The origins of the disease remain conjectural, but feeding young calves with meat-and-bone meals rendered from sheep may be a most likely explanation (Smith and Bradley 2003). Its first incidence was noted in the UK in 1986, and it was eventually found in more than 30 countries (mostly in Europe, also in Canada, the US and Japan) and led to prolonged disruptions of international beef trade. Fortunately, the initial fears about the extent of transmission to humans in the form of vCJD proved exaggerated, and the British statistics show 122 confirmed and another 54 probable vCJD deaths between 1990 and the end of 2011, with the peak of 28 cases in the year 2000 followed by a rapid decline to 5 deaths by 2005 (NCJDRSU 2012).

And the concerns about BSE overlapped with worries about a wide spread diffusion of poultry-borne influenza. A new influenza subtype H5N1, capable of killing nearly all affected chicken within a few days, was first detected in Hong Kong's poultry markets in April 1997, and the very next year this highly pathogenic form caused the first human death, infecting a three-year-old boy directly without passing through an intermediate host (Sims et al. 2003). By the time that episode ended, 18 people died and 1.6 million birds were slaughtered (Snacken et al. 1999). By 2003, a highly pathogenic subtype H5N1 reappeared, and within three years it spread to both domestic and wild birds throughout East and Southeast Asia, and from there it spread westward all the way to several European countries. Fortunately, the strain was not easily transmissible to humans, and by 2005 there were fewer than 100 deaths in Vietnam, Thailand and Indonesia, but the outbreak forced mass slaughter of infected poultry: 40 million chicken were killed in Thailand in 2004 (Chotpitayasunondh et al. 2004).

This virus will always be with us, spreading from its natural reservoirs in South China's duck flocks (Chen et al. 2004) – and it will also retain its pandemic potential (Li et al. 2004). As a result, we cannot exclude the possibility that a future pandemic influenza (whose timing cannot be predicted but whose return is inevitable) will emanate from domesticated poultry. Between 2003 and 2011, avian influenza killed 343 of the 582 infected people, a very high mortality rate of 59% – and a clear cause for concern should the virus become easily transmissible between humans. In the early months of 2012, the virus again killed a small number of people in China and Southeast Asia. Although it is highly unlikely that high (59%) mortality rate based on the known cases of Asian deaths would not be the norm should H5N1 cause a true pandemic, even mortalities on the order of 1–2% would be enough to cause global death rate higher than in 1918 flu pandemic (Butler 2012). And there are always new concerns: in 2011, a new virus (named Schmallenberg after the German town where it was first found) causing fetal malformations and stillbirths began to spread among ruminant animals in Germany, the Netherlands and Belgium; fortunately, its spread to humans appears unlikely (SMC 2012).

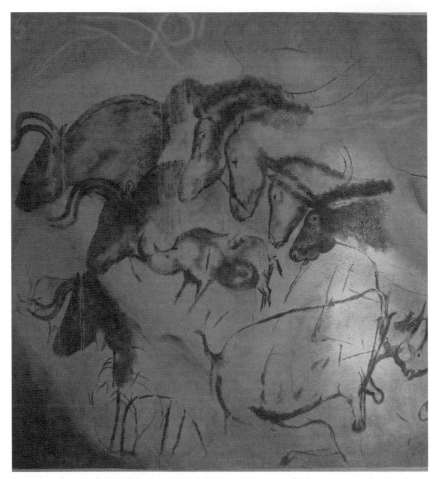

Images of megafauna drawn in charcoal on the walls of Chauvet cave in southern France about 30,000 years ago: these accurate and dynamic pictures make it clear that our ancestors spent much time observing the animals they hunted. The public domain image is available at http://upload.wikimedia.org/wikipedia/commons/d/d1/Chauvet_cave%2C_paintings.JPG

2

Meat in Human Evolution

Eating meat is a part of our evolutionary heritage as much as are large brains (they might actually be that large partly because of meat eating), bipedalism and symbolic language: this is a statement of fact, not a comparison forced in order to begin a book about carnivory. Our hominin ancestors were omnivorous, and, merely in order to survive, our not so distant ancestors inhabiting the periglacial landscapes of the last Ice Age had to be aggressively carnivorous. There is much historical evidence that meat consumption has been a sign of affluence and success and that meat sharing (practiced by chimpanzees) has continued to create personal bonds in most cultures. For centuries, meat was the indispensable foodstuff to energize marching armies, and recently it has been elevated to the cornerstone of voguish high-protein diets. At the same time, meat eating has not only been a symbol of satiety and contentment but also an object of taboos, disapproval and outright scorn, abhorred and renounced by ascetics around the world, be they Dominican *fratres* or Brahmin *sadhus*.

The origins of meat eating – the timing of its emergence, mode of meat procurement, relative amounts of different species eaten – cannot be teased out from the sparse fossil record. Collecting of plant foods presented many opportunities for picking up insects (high in protein) and their larvae (high in fat) and finding eggs and fledgling birds, small rodents, lizards and turtles, while many ancient massive shell middens testify to regular consumption of mollusc and crustaceans gathered

Should We Eat Meat?: Evolution and Consequences of Modern Carnivory,
First Edition. Vaclav Smil.
© 2013 John Wiley & Sons, Ltd. Published 2013 by John Wiley & Sons, Ltd.

along seashores. But reconstructing hominin diets of 2 or 1 million years ago is quite impossible, and we cannot quantify the shares of food energy and dietary protein originating in meat even for the relatively recent past of the Upper Paleolithic (50,000–10,000 years ago) or the Neolithic that had immediately followed it. That has not prevented Cordain (2012) from offering in his *Paleo Diet Cookbook: More than 150 Recipes for Paleo Breakfasts, Lunches, Dinners and Beverages* (not to be taken seriously: how did those hunters on Eurasian and North American plains obtain tomatoes to cook with meat or cinnamon to sprinkle on their baked apple?).

Meat consumption of those pre-agricultural Neolithic hunting societies that did not hunt megafauna cannot be derived from a relatively abundant archaeological record because eating of small animals does not leave reliable traces. But analyses of a stable isotope of nitrogen present in human bone collagen can prove incontrovertibly either a marginal or a significant presence of animal foods in the diets of those periods. And because no milch animals were domesticated before the Neolithic era, animal foods consumed until the latest Paleolithic had to be meat and animals' internal organs. And the quantifications of average or typical diets during the antiquity, Middle Ages and the early modern era (post 1500) are not that much easier.

There are many, often very detailed, descriptions of foodstuffs, meals, culinary habits, dietary preference and eating taboos, and they generally grow more frequent with time. But this textual abundance is over-whelmingly of local, anecdotal, sporadic and qualitative nature, and it does not provide adequate, continuous information needed to calculate typical or modal food intakes, their partition among the principal food categories and nutrients, and their long-term changes. Written sources are thus good enough, particularly when they are combined with demo-graphic and anthropometric data, to offer fairly reliable qualitative judgments that indicate relative levels of food intake and its composi-tion (either as longitudinal comparisons for a region or a small country or as contrast with other societies or cultures during specific periods) and their secular trends.

These accounts make it clear that in many agricultural societies, meat consumption fell, chronically or episodically, to very low levels, be it due to poor productivity of traditional farming, mismatch between rela-tively high yields and high population density in some regions, religiously motivated proscriptions of meat eating, or recurrent food shortages and famines caused by wars and natural catastrophes. But both in global and regional terms, strictly vegetarian communities and societies were always in minority; in most cultures, meat had a status of prestige food, and the desire to consume meat had, with the rarest exceptions, remained strong throughout history and across all economic strata.

Its expression can be found in continuous hunting of wild animals in societies where the eating of domesticated species was forbidden or in the creation of ingenious dishes using other ingredients as meat substitutes. And as soon as greater opportunities to eat meat had presented themselves, during the era of urbanization and industrialization, average per capita meat consumption began to rise, first during the 19th century in North America, Western Europe, Australia and New Zealand. This trend continued during the 20th century when it was transformed into a great global dietary transition that has seen meat consumption reaching saturation rates in some Western countries and rising to unprecedented levels in many populous and rapidly modernizing nations in Latin America and Asia.

Hunting Wild Animals: Meat in Human Evolution

Species belonging to our genus – going back to *Homo habilis* who first appeared about 2.3 million years ago – had spent more than 99.5% of their evolution as foragers, and our species, *Homo sapiens*, has led the same kind of existence for no less than 95% of its development. All of these hominins shared an essentially identical repertoire of survival choices: gathering any edible phytomass (including starchy tubers, fibrous stalks, soft leaves, and nutritious fruits, nuts and grains), collecting insects (and their larvae) and terrestrial and aquatic invertebrates (snails, bivalves), snaring or simply catching small mammals, and using tools to hunt large herbivores (first with spears and much later with bows and arrows) and to catch fish and aquatic mammals (with spears, nets and baskets).

While there is no doubt that meat eating has had an important role in human physical and behavioral evolution, we can only speculate about when, how and why humans began to eat substantial amounts of meat (Stanford and Bunn 2001). All hominins were omnivorous, but the environments they lived in and their hunting skills dictated the prevailing levels of meat consumption, with the lowest rates in the tropical rain forests and the highest rates on grasslands. For some groups, meat remained a marginal foodstuff, but for many of them it supplied a significant portion of their food energy and an even larger share of their dietary protein, and many late Paleolithic and Neolithic hunters in boreal and sub-boreal environments were predominantly carnivorous. Those hunters may not have been entirely responsible for the relatively rapid extinction of Pleistocene megafauna, but there is no doubt that their actions had contributed to the demise of the largest herbivores in Europe, North America, Australia and large parts of Asia.

The long duration of these foraging experiences and the enormous variety of environments they had eventually dispersed to from their ancestral African lands (living in arid grasslands as well as in the richest forests, in climates from the tropics to the sub-Arctic) led the foragers to develop a remarkable range of hunting strategies, ranging from simple snares for catching small animals to group hunts relying on elaborate ruses to stampede entire herds, and from using well-crafted tools to deploying other animals (cormorants, birds of prey, dogs) as efficient hunters. Meat, internal organs and fat secured by these hunts made not only an important contribution to the physical evolution of hominins (above all to their smaller gastrointestinal tracts and to their very high encephalization) but also to their social development: meat is, indisputably, one of the key defining components of our evolutionary heritage. And hunting did not cease in the settled agricultural societies; again, in some of them it was reduced to a marginal role while in others the practice had persevered for millennia and was a welcome source of additional high-quality protein and fat.

While a relatively small number of mammalian and avian species dominates the modern Western commercial selection, historical and global diversity of meat eating has always been an all-encompassing affair: organisms of every size (from maggots to mammoths) and every vertebrate and invertebrate order have been collected or killed for meat. A complete list of muscle and associated fat tissues consumed in the past would thus contain thousands of wild species, and even if a list were limited to species that are habitually consumed by large numbers of people (at least tens of thousands, often tens of millions), it would contain some of the smallest as well as some of the largest organisms (from small songbirds weighing less than 100 g to large bulls whose body mass surpasses 1t).

Ill repute in one culture is no obstacle in another: for example, grass rats (*Arvicanthis*) and giant rats (*Cricetomys*) are cherished foods in large parts of West Africa (Njiforti 1996). Cuteness and small size offer no protection: domesticated guinea pigs become temporary pets around the kitchens of Peruvian houses before they are roasted (Charbonneau 1988). Massacres on grand scale are an accepted way of getting meat: although the practice is illegal, the Cypriots have been using mist nets to trap and kill millions of migrating songbirds every year, discarding the unwanted ones and grilling and pickling the prized ones (Bird Life International 2011). Nor does being near extinction offer any protection for a hunted species: wild-meat markets in Congo offer seared body parts of nearly extinct mountain gorillas and chimpanzees.

Primates and hominins

Foraging and scavenging habits of the earliest hominins had to bear some similarity to the food acquisition of their primate ancestors (Whiten and Widdowson 1992). Long-term field studies of chimpanzees, our closest primate ancestors, in several locations in Central and Western Africa have shown that both the common chimpanzees (*Pan troglodytes*) and the less studied smaller bonobos (*Pan paniscus*) are fairly regular hunters and consumers of meat. Chimpanzee hunting targets mostly small monkeys (particularly red colobus), while bonobos were observed hunting duikers, flying squirrels and some small primates (Boesch 1994; Stanford 1999; Boesch and Boesch-Achermann 2000; Hohmann and Furth 2008). Average consumption of meat among the studied chimpanzee groups has been between 4 and 11 kg/year/capita, as much or more than in many traditional agricultural societies of the last millennium.

Observations of chimpanzee hunts have also shown some cooperative features and subsequent meat sharing by successful males. These practices maximize an individual's chance of getting meat's high-quality protein and essential micronutrients (Tennie et al. 2009). Moreover, meat sharing also reinforces social bonds within a group, and the exchange of meat for sex appears to be done by males on a long-term basis (Teleki 1973; Boesch 1994; Stanford 1999; Gomes and Boesch 2009). This does not mean that occasional cooperative hunting by chimpanzees should be seen as a prototype of later human behavior: an opportunistic arm-swinging pursuit in canopies has little to do with planned bipedal hunting with weapons. I agree with Sayers and Lovejoy (2008) who argue that the chimpanzee referential models for early hominin behavior have been too often misapplied (and that only differences between closely related forms explain their divergence), but the fact our closest primate ancestor is also a hunting omnivore helps in understanding the similar features in the evolution of our species.

Evidence for the hominin and early human omnivory is rich and indisputable (Milton 1999, 2003; Kiple 2000; Larsen 2000, 2003; Richards 2009). No foragers, and particularly not those living in environments subject to pronounced seasonal fluctuations in semi-arid or boreal environments, could ignore animal foods. Hominin consumption of animals in Africa began most likely with collecting invertebrates (insects, snails, molluscs) and catching amphibians and small mammals and with scavenging carcasses of larger animals (Domínguez-Rodrigo 2002; Domínguez-Rodrigo and Pickering 2003). Because of their limited capabilities (the early humans being smaller and less muscled than modern adults) and the absence of effective weapons, many anthropologists have believed that our earliest ancestors were much better scavengers than hunters (Binford 1981; Shipman 1986; Blumenschine and Cavallo 1992).

Large predators (above all lions and leopards) often left behind partially eaten carcasses that could be reached by alert and enterprising hominins before they were claimed by vultures and hyenas. Binford (1981) was the main proponent of an "obligate marginal scavenger" model according to which the foragers of the Lower and Middle Paleolithic were passive scavengers of carnivore-killed carcasses. Blumenschine (1991) argued that the hominins concentrated on nutritious bone marrow that remained in ungulate limbs after their defleshing by carnivores and Stiner (1994) favored a model that combined opportunistic scavenging with hunting. A major argument against seeing early hominins as frequent scavengers has been the fact that meat-eating primates almost always avoid carrion even if the kill is fairly fresh because its eating is likely to result in a gastrointestinal illness (Ragir et al. 2000).

But incontrovertible evidence of cut marks over carnivore tooth marks on herbivore bones shows that some scavenging had to be taking place, while contemporaneous presence of highly cut-marked long limb shafts indicates active hunting of small- and medium-sized animals (Domínguez-Rodrigo et al. 2009; Blascoa et al. 2010). Long-lasting anthropological disputes about the relative importance of passive or aggressive scavenging and hunting can be largely closed for a specific site (e.g., see Villa et al. 2005), but there can be no generally valid conclusion applicable to different environments and to periods ranging over more than two million years. And it is important to note that scavenging and hunting with weapons were not the only options for more evolved hominins. In open landscapes, persistence hunting, or more accurately running down a prey in long-distance pursuits, was actually an efficient way of bringing down some fairly large animals (Liebenberg 2006).

Whatever the means of getting meat, meat eating among the early hominins (those predating *Homo erectus*) has been clearly documented in the Koobi Fora Formation in East Turkana (northern Kenya) thanks to a unique combination of faunal remains (both terrestrial and aquatic animals, including pigs, turtles and hippos), Oldowan artifacts dated reliably to 1.95 million years ago and some direct evidence of butchery (Braun et al. 2010). Archaeological evidence suggests that meat eating was on the increase some 1.5 million years ago, and eventually meat scavenged from the kills by large carnivores began to be augmented by deliberate hunting. Increased frequency of meat eating helps to explain a roughly 40–50% gain in both height and body mass that took place between the appearance of *H. habilis* (about 2.3 million years ago) and *H. erectus* about half a million years later (McHenry and Coffing 2000), as well as reduced size of jaws and teeth in the latter species (Aiello and Wells 2002).

But carnivory's impact has gone much beyond a supply of high-quality substitute for plant foods. Stanford's (1999) observations of chimpanzees led him to link the origins of human intelligence to meat not because of nutritional qualities but because of the refinement of cognitive abilities required for the strategic sharing of the meat within the group: "The intellect required to be a clever, strategic, and mindful sharer of meat is the essential recipe that led to the expansion of the human brain." While primates and early hominins have an average encephalization quotient (actual/expected brain mass for body weight) between 2 and 3.5, its value for humans is slightly above 6 (Foley and Lee 1991). Three million years ago, *Australopithecus afarensis* had brain volume less than 500 cm^3, 1.5 million years ago *H. erectus* had a nearly doubled brain size and adult brains of *H. sapiens* are roughly still 50% larger (Leonard et al. 2007).

This brain growth was directly linked with meat consumption by the expensive-tissue hypothesis formulated by Aiello and Wheeler (1995) in order to explain how the very high metabolic cost of much larger brains was met without a corresponding increase in overall basal metabolic rate. Per unit of mass, brain needs about 16 times as much energy as does skeletal muscle, and at about 350 g, neonate brain is twice as large as that of a newborn chimpanzee, and it becomes more than three times as massive as the brain of our closest primate species by the age of five (Foley and Lee 1991). Because humans do not have more of metabolically expensive tissues (internal organs and muscles) than would be expected for a primate of our size, Aiello and Wheeler (1995) argued that the only way to support larger brains without raising the overall metabolic rate was to reduce the size of another major metabolic organ.

With little room left to reduce the mass of liver, heart and kidneys, the gastrointestinal tract was the only metabolically expensive tissue whose size could vary quite a bit depending on the dominant diet. Diet of higher quality would allow the reduction in relative gut mass – Chivers and Hladik (1980) showed that the shape and size of mammalian gastrointestinal tracts are related to diet quality – and could be the primary factor in the evolution of larger brains. Fish and Lockwood (2003, 171) found that diet quality and brain mass have a significantly positive correlation in primates (their analysis included more than 40 species) and that, consistent with the expensive-tissue hypothesis, better nutrition allowing the reduction of the relative gut mass "is one mechanism involved in increased encephalization."

Leonard et al. (2007) analyzed data on resting metabolic rate, brain : body size and diet quality for humans and 40 primates and made revealing comparative conclusions: that for a given metabolic rate, primate brain sizes are about three times those of other mammals, and human brain size is three times that of other primates; that humans have an extreme combination of large brain : body size and of a substantially higher-quality diet than is indicated by their size; and that these better diets including meat have, in part, supported those larger brains whose high metabolic cost has been partially offset by reduced gastrointestinal tract. And Milton (1999, 2003) supported the argument about the importance of substantial carnivory by looking at the evolution of the gastrointestinal tract.

Because hominins came from an ancestral lineage whose survival was dependent on plant foods, the human gut is very similar to that of extant primates with one notable exception: while all apes have more than 45% of their gut in the colon and only 14–29% in the small intestine, the proportions are reversed in humans, with more than 56% in the small intestine and only 17–25% in the colon. This clearly indicates an adaptation to high-quality, energy-dense foods that can be digested in the small intestine, that is, to a diet containing some meat, including fat and the nutrient-rich internal organs. Their inclusion in the hominin diet made it easier to lead a highly active, mobile existence in places where the strategy used by chimpanzees (getting most of their energy and protein from nutritious fruits) was impossible – while the feeding strategy adopted by gorillas and orangutans would severely limit activity and mobility as it requires massive, mostly sedentary, daily intakes of low-quality plant matter.

The expensive-tissue hypothesis was questioned by Navarette et al. (2011) who concluded, after testing a sample of 100 mammalian species (including 23 primates), that the brain size is not negatively correlated with the mass of the digestive tract but with fat reserves. But this claim rests on the correlation between residual brain mass and residual adipose deposits in 28 wild-caught female mammals whose r^2 is merely 0.258 (leaving most of the relationship unexplained), and this weak link would be further weakened after excluding the observations for Chiroptera (bats). I see this analysis as an indicator of (not unexpected) evolutionary complexity but not, as the authors claim, a refutation of the expensive-tissue hypothesis. Moreover, this claim ignores the undoubted uniqueness of human evolution. While we cannot resolve the disputes about the precise contribution made by higher meat intakes to encephalization in hominins or to their role in the diets of weaned children, there can be no doubt that this dietary shift was of enormous importance in human evolution (Milton 1999; Mann 2007).

Meat consumption during the Paleolithic period

Zooarchaeological evidence makes it clear that the Neanderthals (*Homo neanderthalensis*) living in most of Europe and the Middle East during the Middle Paleolithic (300,000–30,000 years ago) were deliberate, rather than just opportunistic, hunters and that the same was true about the anatomically modern humans that displaced them during the Upper Paleolithic (Drucker and Bocherens 2004). Our species (*H. sapiens*) was first identified in Africa about 195,000 years ago, and its early hunting capabilities were no more advanced than those of its predecessors (Trinkhaus 2005; Klein 2009; Rightmire 2009). Larger mammals could be hunted only once weapons with a longer reach became available. The earliest preserved throwing (javelin-type) spears, enabling attacks from a greater distance, were found in Germany in 1996: six spears, 2.25 m long and made from a spruce trunk, date as far back as 380,000–400,000 years ago (Thieme 1997).

These weapons are an incontrovertible proof of planned hunting, but we can only speculate if they were used only by individuals or in cooperative hunts. Hunting of large animals (aurochs, horses, wild pigs and asses) and the subsequent sharing of their meat was also convincingly demonstrated by tool marks on animal bones from the Qesem Cave in Israel perhaps as far back as 400,000 years ago (Stiner et al. 2009). Only about 50,000 years ago did humans begin to make a variety of specialized tools, using not only stone but also bone, ivory and antlers, and the first sturdy bows and arrows came about only 25,000 years ago. Better tools were one of the advantages that helped humans displace the Neanderthals, colonize the northern latitudes and create some remarkable cave drawings and sculptures.

While the Neanderthal diet was that of a top carnivore (Richards 2002), the Paleolithic human diets combined animal and plant sources, with the former necessarily more prominent in the mid-latitudes during the last glacial maximum (around 20,000 years ago, with the southernmost ice fronts at about 50°N in Europe). Judging by typical skeletal remains, these diets were generally adequate, but because plant remains and organic food residues rarely survive, their specific reconstruction is impossible. Analysis of stable carbon and nitrogen isotopes in preserved bone collagen is the best available proxy for the qualitative evaluation of ancient diets (Drucker and Bocherens 2004).

Humans eating only plant protein would have $\delta^{15}N$ values identical to those of herbivores living in the same region, those eating only animal herbivore protein would have $\delta^{15}N$ value equal to that of local herbivores plus an enrichment value of 3–6‰, and eating mixed protein would result in linear enrichment ratios between those two extremes. A number of practical difficulties restrict the use of this technique to an indicative (qualitative) level (Hedges and Reynard 2007), but there can be no doubt that this isotope evidence shows the presence of enrichment, and hence of significant meat consumption, in every study of late Paleolithic and Neolithic bone collagen.

Recently published examples for collagen from the late Upper Paleolithic include terrestrial herbivores as the main source of protein for the foragers in northern Spain (Garcia-Guixe et al. 2009) and in Cosenza in southern Italy (Craig et al. 2010), and a combination of terrestrial and marine mammals in Wales (Richards et al. 2005). Archaeological evidence from such famous Upper Paleolithic sites as Solutré in eastern France – with the remains of at least 32,000, and perhaps 100,000, horses that were driven up from the valley floor during summer migrations into a cul-de-sac formed by cliffs, killed by spears and butchered over a period of some 20,000 years (Olsen 1989) – make it clear that some groups derived most of their food energy from meat. A general conclusion is that during the late Paleolithic period, meat of herbivorous animals was an important component of diets in large parts of Europe (Holt and Formicola 2008), but that an important change began during that time thanks to the expansion of diet diversity, above all to a greater consumption of aquatic species including mammals, water fowl and invertebrates.

Finch and Stanford (2004) argued that the dietary shift to regular consumption of fatty meat in hunting societies was mediated by the selection for "meat-adaptive" genes that conferred resistance to risks associated with meat eating while increasing longevity of regular meat eaters. Paleolithic meat consumption was made more digestible, more palatable and safer by widespread cooking (searing, roasting, smoking), whose invention and adoption had, according to Wrangham (2009), a monstrous effect on human evolution. While the earliest consumption of some cooked food might have taken place as far back as 1.9 million years ago, the earliest indisputable dating of hominin use of fire was pushed from about 250,000 years ago (Goudsblom 1992) to 790,000 years ago (Goren-Inbar et al. 2004). But the common use of fire goes only to the Middle Paleolithic (300,000–20,000 years ago), the period when *Homo sapiens sapiens* displaced the European Neanderthals (Bar-Yosef 2002; Karkanas et al. 2007).

Paucity of direct evidence regarding the composition of specific Paleolithic diets has not stopped a great deal of speculation about their actual makeup. Perhaps most notably, Eaton and Konner (1985) and Eaton et al. (1997) used information on diet of foraging societies that had survived into the 20th century to estimate that Paleolithic hunters were deriving 34% of all food energy from protein. Given the fact that they ate no cereals and only protein-poor fruits and tubers, nearly all of it had to come from animal foods, mostly from meat of hunted herbivores. By 1997, they raised this share to 37% (Eaton et al. 1997), and Cordain et al. (2000) estimate the share of food energy from animals at more than 60%, a very high level compared to the recent US estimate of less than 30% (USDA 2012a). The authors were interested in reconstructing the ancient diet mainly in order to offer their discordance hypothesis according to which the shift away from meat to grain intakes (as well as reduced activity) "have contributed greatly and in specifically definable ways to the endemic chronic diseases of modern civilization" (Konner and Eaton 2011, 594).

The great range of Paleolithic environments precluded any universal Paleolithic diet but – assuming that only a third of all food energy came from plant foods and the rest from terrestrial and aquatic animals – its specific features are clear: very high intakes of animal protein, fiber and (in warmer climate) fruits, medium to high intakes of fat and cholesterol, very low or no intake of cereals and tubers, and no dairy products; as for the micronutrients, it was very high in virtually all minerals (except for low intake of sodium) and nearly all vitamins (Eaton and Eaton 2000). The most remarkable, and paradoxical, attribute of these diets from the modern point of view was that they were meat-based yet non-atherogenic (Cordain et al. 2002), a fact explained by a relatively low fat content of many wild meats and by their (compared to the meat of domesticated animals) higher shares of mono- and polyunsaturated fatty acids. Consequently, the modern risk is due to excessive intake of saturated fat associated with the meat of domesticated livestock, not with meat itself (Mann 2000).

Given the small numbers of Paleolithic hunters – the best estimates of global population are no higher than a few hundred thousand people 50,000 years ago and perhaps as many as 2 million by the end of the latest glaciation – their impact on the total numbers and typical densities of hunted animals had remained negligible for millennia. This may have changed little as the numbers of periglacial hunters increased with the receding ice fronts, but an intriguing hypothesis first formulated in the middle of the 19th century by Owen (1861) suggested otherwise: is it not possible that the human quest for fatty meat of large herbivores was a major, if not the principal, reason for a relatively rapid extinction of the late Pleistocene megafauna?

Extinction of the late Pleistocene megafauna

When applied to the late Pleistocene extinctions (the epoch ended 12,000 years ago), the term *megafauna* refers not only to mammoths and other massive herbivores but also to all mammals with adult body exceeding 40–50 kg, that is, the mass of a young deer. The largest species in Eurasia were woolly mammoth (*Mammuthus primigenius*), woolly rhinoceros (*Coelodonta antiquitatis*) and common large herbivores, which included giant deer (*Megaloceros giganteus*), the steppe bison (*Bison priscus*) and auroch (*Bos primigenius*). In North America, there were also mastodons (*Mammut americanum*) and giant ground sloths (*Megatherium*), while Australia had giant herbivorous marsupials.

Megaherbivores make attractive hunting targets not only because a kill is rewarded with a large mass of meat (yielding large multiples of food energy spent on the hunt) but also because they contain much more fat than do smaller species, and hence their energy density may be more than twice as high as that of animals weighing only a few kilograms. Killing a single woolly mammoth could thus yield as much edible energy as killing 100 large deer and killing a small bison was equal to killing more than 200 rabbits. Group hunting of these animals commonly yielded up to 50 times as much food energy as was invested in their killing, leaving only near-shore whaling as a more energetically rewarding activity (Smil 2008).

Moreover, some megaherbivores could be killed without any weapons: well-planned and skillfully executed stampeding of buffalo herds over cliffs – commemorated in Alberta's Head-Smashed-In Buffalo Jump near Fort Macleod – is perhaps the most impressive example of this ingenious hunt (Frison 1987).

Modern revival of Owen's (1861) extinction-by-hunting hypothesis began during the late 1950s with Martin's publications, whose flow continued for the next five decades (Martin 1958, 1967, 1990, 2005). Martin's often cited "overkill hypothesis" attributed a "blitzkrieg" extinction of North American megafauna to an advancing wave of hunters migrating from Beringia southward all the way to Tierra del Fuego. Radiocarbon dating of animal tissues boosted the hypothesis by constraining much of the extinction into relatively brief time spans coinciding with the arrival of hunters in Australia and in North America. According to Roberts et al. (2001), Australia's animals with body mass heavier than 100 kg became extinct around 46,400 years ago, or within just 1,000 years of human arrival, and nearly half of North America's late Pleistocene extinctions were concentrated between 12,000 and 10,000 years ago, immediately after the arrival of the Clovis people, the first hunters from Asia, between 13,200 and 12,800 years ago (Thomas et al. 2008).

Slow breeding rates among large mammals (long gestation periods and birth of single offspring) and their persistent hunting can be modeled to produce rapid population declines. Even when assuming slow human population growth rates, random hunting and low maximum hunting effort, Alroy's (2001) simulation of the Pleistocene–Holocene extinction in North America indicated median occurrence in only about 1,200 years after the initial invasion by humans, a time span close to that between the earliest Clovis artifacts and the extinction of the North American megafauna. But the same assumptions do not work in Africa, where megafauna was hunted long before any humans reached North America and where no large herbivores went extinct for another ten millennia.

Looking at individual species in specific regions offers different perspectives. Some large species became extinct well before the late Pleistocene (straight-tusked elephant), others have been under stress due to climate changes, in Euroasia as well as in Australia (Stuart 2005; Pushkina and Raia 2007; Webb 2008). Perhaps most notably, there is no convincing evidence of any rapid extermination of woolly mammoths, Euroasia's largest herbivore. Hunters and mammoths had coexisted for millennia, and the prime reason for mammoth extinction were the climatic changes during the late Paleocene and early Holocene that led to the replacement of nearly treeless periglacial vegetation capable of supporting masses of large herbivores by tree-dominated ecosystems (Reumer 2007; Stuart and Lister 2007; Kuzmin 2009; Allen 2010).

Moreover, some megaherbivores, including those that make perfect hunting targets, survived well into the Holocene. Giant deer (Irish elk, *M. giganteus*) lived in western Siberia until about 7,700 years ago (Stuart and Lister 2007), and in Alaska and Yukon, bison (*B. priscus*) and wapiti (*Cervus canadensis*) became actually more numerous both before and during the human colonization of the region (Guthrie 2006, 209). And even some mammoths made it into the Holocene in at least two isolated refugia in Siberia and on the Wrangel Island in the Arctic where diminutive mammoths lived as recently as 2,038 BCE (Vartanyan et al. 1995), half a millennium after the pyramids were built.

More than 50 years after Martin opened the modern extinction debate, there is no real consensus about the causes of megafauna demise, but the majority opinion now ascribes them to a combination of natural factors and human actions and not just to determined slaughter by carnivorous hunters (Koch and Barnosky 2006; Yule 2009). Climate change and the ensuing transformation of plant cover appear to be the key factor, and human actions, mainly selective hunting and anthropogenic fires, had a contributory effect.

I have tried to estimate the minimum aggregate number of megaherbivores killed every year during the late Pleistocene (Smil 2013). Such a calculation cannot aspire to accuracy greater than the right order of magnitude. I assumed the late Pleistocene population of at least two million people and high average per capita food requirements of 10 MJ/day due to their high mobility and life in cold climate. The average per capita rate (including all children and adults) could not have been higher than about 10 MJ/day. Mosimann and Martin (1975) assumed that the late Pleistocene foragers ate mostly meat and that about 80% of it came from megaherbivores and the rest mostly from smaller species, with the live weight of killed animals about 2.5 times the weight of their edible tissues whose average energy density was as much as 10 MJ/kg.

Given these assumptions, the late Pleistocene hunters would have to kill nearly 2 Mt (fresh weight) of megaherbivores every year. The minimum number of slaughtered animals can be estimated by assuming that all of those kills were mammoths; depending on the shares of actually killed *M. americanum*, *M. primigenius* and *Mammuthus meridionalis* weighing 4–9 t, and of a larger *Mammuthus imperator* weighing more than 10 t (Christiansen 2004), this would translate to no fewer than 250,000 and to as many as 400,000 animals. A more realistic assumption of hunting a mixture of megaherbivore species with an average body mass on the order of 1 t/animal would raise the annual global kill to roughly two million large mammals.

But even these order of magnitude estimates are uncertain. We do not know the meat/plant shares of regional Pleistocene diets, and even with a highly meaty diet a significant part of food energy could have come from small animals and the total number of large mammals and the body mass distribution of kills could have been bimodal, including a larger number of smaller animals (less than 50 kg/head), not that much in between and a smaller number of the largest (in excess of 1 t/head) species. As a result, the total number of herbivores annually killed by the late Pleistocene hunters could have easily differed by ±50% from the estimates. In any case, an annual kill of some 2 Mt of live weight would have been an impressive total: the recent annual slaughter of cattle, the world's dominant domesticated megaherbivores, has been about 300 million animals annually or just over 100 Mt of fresh weight – but supplying meat for a population larger than 6 billion!

Hunting in different ecosystems

The most fundamental variable that determines the density of zoomass that is potentially available for hunting is the magnitude of energy transfer between trophic levels. Shares of the net photosynthetic productivity consumed by herbivores are only about 1–2% in temperate forests, may be higher than 25% in temperate grasslands and they reach global maxima of 50–60% in rich tropical grasslands (Crawley 1983; Chapin et al. 2002). Even though Hairston et al. (1960) showed that the numbers of herbivores are usually limited by predators rather than by energy available for transfer, abundance of grazers on highly productive grasslands meant that, once the improved techniques made it possible to kill large herbivores in numbers sufficient for group survival, grasslands became the optimum hunting habitats.

In contrast, the world's most biodiverse ecosystems were inferior environments for highly rewarding hunts. Most of the mammals living in tropical rain forests are folivorous and hence arboreal, relatively small and difficult to locate in high tree canopies, and many of them are also nocturnal. As a result, bow-and-arrow hunters in tropical rain forests had always faced much lower chance of hunting success than their counterparts in open woodlands or grasslands, and the meat of small mammals may have returned as little as two to three times as much energy as invested in the pursuit, with many hunts ending with no kills.

Sillitoe's (2002) studies in a tropical rain forest of the Papua New Guinea highlands found that hunters expended up to four times more energy on hunting than they gained in meat and fat and concluded that the survival in those environments required not only diligent plant gathering but also some proto-horticultural practices. Long before these studies, Bailey et al. (1989) generalized the rule by thinking it unlikely that any foragers could have ever survived in any tropical rain forest as pure hunters without any access to cultivated foods. New archaeological evidence from Malaysia led them to modify their original conclusion as they conceded that such an existence was possibly in Asian tropical rain forests with high densities of sago palm and with relatively high densities of wild pigs (Bailey and Headland 1991). For other tropical environments, their original conclusion appears to be justified.

Where a zoomass-rich environment presented many hunting opportunities, it was the average body mass of prey that became the key determinant in choosing the targets. The smallest herbivores (rodents, lagomorphs) offered usually the least rewarding pursuits, not only because they yielded so little meat but also because they so easily evaded clubs or arrows or simple snares. Killing the largest herbivores brought the highest energetic payoff, but such animals reproduced and matured slowly and their hunting was also more dangerous as they could easily kill a careless hunter. As a result, the most commonly hunted animals combined a fair body mass with relatively high reproductive capacity and a high territorial density.

Rapidly reproducing wild pigs, whose adult body mass was up to 90 kg, were a favorite target in both tropical and temperate environments, as were deer and antelopes, with the smallest animals weighing less than 25 kg and the largest one well over 500 kg. But the most desirable targets were large ungulates, not only because of a large meat package on hooves but also because of their above-average fat content. Hunters everywhere had to reckon with the fact that edible tissues of most small- to medium-size hunted animals had a relatively low food energy density. Fresh carcasses of animals ranging from lagomorphs (hares and rabbits) to smaller ungulates (deer, elk, smaller antelopes) have typically more than 70% of water, 20–23% of protein and less than 10% (and in some species less than 1%) of subcutaneous and intestinal fat. Their meat is virtually pure protein with food energy density ranging mostly between 5 and 6 MJ/kg, or only 30–40% energy density of carbohydrate staples averaging 15 MJ/kg, and less than 5% of fat.

Some smaller but fatty underground mammals, including antbears and porcupines, are the only exception: they were sought after despite much labor needed to dig them out from their deep burrows. Leanness of wild meat explains why even abundance of small herbivores could leave hunters craving to eat meat with higher food energy density (i.e., with a substantial share of satiety-inducing lipids). Prolonged feeding on nothing but lean meat causes acute malnutrition known as rabbit starvation (or *mal de caribou* of the French explorers) that brings nausea, diarrhea and eventual death (Speth and Spielmann 1983). Hayden (1981) took this implication to its extreme conclusion by claiming that hunters' real quest was for fat and that the idea of meat being held in high regard by hunting societies is a misconception created by ethnographers. This is arguable but the preference of all hunters for killing fatty animals is indisputable. That is why megaherbivores were always preferred, as greater (and often fatal) risks associated with their hunting were trumped by superior energy returns.

As already detailed, Pleistocene hunters went after woolly mammoths, animals with body mass two orders of magnitude larger than their bodies, and after other massive herbivores (elephants, rhinos and aurochs), which is why the Kalahari San preferred to kill elands, the largest of all antelopes (body mass in excess of 600 kg), and why North American Indians favored bison. And high intakes of lipids were indispensable in the Arctic where energy requirements for basal metabolism and strenuous activities in the cold environment could be met only by consumption of fatty marine mammals (Cachel 1997). High fat content also explains why energy returns in hunting large terrestrial herbivores were commonly two or three times higher than those for hunting leaner species weighing less than 100 kg, five to six times higher than capturing small (less than 5 kg) animals – and easily ten times higher than digging up starchy roots or collecting sweet fruits (Cordain et al. 2000; Smil 2008). Those high rewards for hunting large terrestrial herbivores were surpassed only by the killing of exceptionally fatty seals and whales.

Hunters deriving large shares of protein from terrestrial megaherbivores or those Arctic groups subsisting almost exclusively by eating fatty marine mammals had no major micronutrient deficiencies because they were also consuming all internal organs of killed animals that are, compared to muscles, high in vitamins C, D and E, as well as in riboflavin, niacin, B12, folate and iron (Hockett and Haws 2003). Lipids of wild animals also have more mono- and polyunsaturated fatty acids than do domesticated species, a fact at least partially explaining the absence of atherogenesis among highly carnivorous hunters.

As for the actual frequency, amount and shares of meat consumption in traditional hunting societies, we have to rely on information gathered by ethnographers and anthropologists studying those groups that endured into the latter half of the 20th century and whose food supply and typical dietary intake could be analyzed by modern scientific methods (numerous first-contact descriptions from the 19th century do not contain reliable quantitative analyses of hunters' diets). This reality means that many, if not most, extant hunting groups were influenced by the contact with neighboring pastoralists or farmers (Solway and Lee 1990). Moreover, some hunting studies that received a great deal of public attention – such as those of the San (Bushmen) of Kalahari (Tanaka 1980; Silberbauer 1981) – dealt with groups living in extreme environments and should not be used to generalize about the shares of food derived in the past from hunting in more equable climates and more fertile non-tropical regions.

An analysis of information on diets in more than 200 foraging groups showed that the share of animal foods obtained by hunting and fishing averaged 65%, and it was still 59% after excluding the Arctic groups that have no choice but to rely almost exclusively on meat, fat and internal organs (Cordain et al. 2002). In those foraging societies whose diets were studied after 1950, the share of food energy derived from animals was also 65%, with the extremes ranging from 26% for G/wi (a society of Kalahari San) to 99% for Alaska's Nunamiut (Kaplan et al. 2000). Cordain et al. (2000) had also categorized the origins by ecosystem and separated hunted and fished foods; terrestrial animals contributed 56–65% in temperate grasslands, 36–45% in tundra but only 16–25% in temperate and 26–35% in tropical forests. Shares of food energy from the hunting of terrestrial animals ranged from about 20% for the coastal tribes in the Pacific Northwest (they derived 50–60% of their food from fishing), to 50–80% among the interior Indian tribes of North America to almost 90% for the Nunamiut.

In terms of average annual per capita meat consumption, these shares translate into a wide range of intakes, often with large seasonal variation (Wadsworth 1984). Kalahari San consumed daily as little as 15 g of meat and as much as 260 g/person, averaging 50–70 kg/year/capita (Tanaka 1980; Silberbauer 1981). In contrast, in the tropical forests of Papua New Guinea, where yams and sago dominated the diet, annual meat intakes were rarely above 25 kg a year, while during the time of abundant kills, adult Inuit in Canada and Greenland could eat 2–4.5 kg meat a day and with high food energy intakes (often in excess of 15 MJ/day), with meat supplying at least 40% of that energy (fat accounted for most of the rest).

As already noted, meat provision also had important social consequences as it required cooperation and sharing. An individual hunter did not have much chance pursuing very large animals, and his daily success rate in hunting smaller mammals was usually less than 15–30%; as a result, at least 3–6 hunters (or a minimum group of 18–40 people) were needed to assure a daily success. A detailed account of Mbuti hunts in the Ituri tropical rain forest of Congo shows that solitary archers averaged just 110–170 g/meat/capita, spear hunters provided 220 g but net hunting yielded 370 g for every person in the group (Harako 1981). And megaherbivore hunts also needed several hunters to butcher an animal and to transport its pieces to a camp.

Anthropologists have spent decades arguing about the reasons why hunters share meat, especially the meat of large animals (Hill and Kaplan 1993). Common explanations have included the reduction of the risk inherent in hunting megaherbivores (i.e., minimizing the chances of hungry days by sharing with other hunters who, too, will share their kills in the future) and altruistic sharing with the hunter's kin. Both may be true, but other factors matter: Hawkes et al. (2001) concluded that meat sharing by Hadza hunters of northern Tanzania does not fit the common expectation of risk-reducing reciprocity and that a hunter's main goal appears to be to enhance his status as a desirable neighbor. Whatever the reasons, meat sharing has been clearly an important means in the evolution of human cooperation and in the enhancement of social cohesion.

Wild meat in sedentary societies

Transition to sedentary societies engaged in regular crop cultivation had to be a gradual process: despite the persistent reference to the term coined by Childe (1951), there was never any abrupt agricultural revolution. Consequently, foraging in general and hunting in particular continued to be an important source of food in all early settled agricultural societies. Excavations at Çatalhöyük – a large Neolithic settlement on the Konya Plain established at about 7,200 BCE and considered one of the earliest examples of a new sedentary civilization – uncovered not only the importance of cultivated staple grains but also the bones of animals (aurochs, foxes, badgers, hares) regularly hunted by some of the world's earliest farmers (Atalay and Hastorf 2006). And in another ancient agricultural settlement in the eastern Mediterranean – in Tell Abu Hureyra in northern Syria – excavations by Legge and Rowley-Conwy (1987) discovered that hunting (above all of gazelles) remained an important source of food for 1,000 years after the beginning of plant domestication.

Hunting remained a common practice in the oldest alluvial civilizations, in ancient Egypt as well as during the early Chinese dynasties (Hartmann

1923; Chang 1977) – in Egypt, the hunted species included ducks, geese, antelopes, wild pigs, crocodiles and elephants; in China, the hunted species ranged from wild pigs to elephants (Elvin 2004) – and in Europe it continued to supplement predominantly plant-based diets of the Roman world, the Middle Ages and the early modern (post-1500) era. The popularity of hunting has also been attested by a large number of still lifes – particularly those depicting such smaller game species as partridges, pheasants, ducks and hares, less frequently wild boars and deer – and by the frequency of poaching in forests.

In all affluent countries, the process of urbanization and the rise of large-scale commercial meat production had relegated hunting of wild animals to a pastime rather than to a supplementary means of meat supply, but some species – including wild boar, deer, hares, grouse, pheasants and partridges – remain a minor (and largely seasonal) source of food in many European countries. In North America, deer, elk, moose and caribou are eaten not only by the natives but also by some rural and urban hunters. In contrast, wild meat is a major source of protein in many parts of Africa that have either no or insufficient large-scale, affordable supply of commercially produced meat by domesticated animals but whose inhabitants have income high enough to afford at least some varieties of much cheaper but often illegally hunted bush meat that is available (fresh, roasted, dried or smoked) in local village and urban markets.

As expected, there are no statistics regarding this meat consumption, and a relatively small number of specific studies refers to quantities that are representative of only particular location and periods and cannot be used for extrapolating nationwide totals. A rare nationwide estimate put Ghana's annual consumption of wild meat at the minimum of 385,000 t during the 1990s (Ntiamoa-Baidu 1998), or a bit over 20 kg/capita. That rate would have been at least twice the total reported by the FAO's food balance sheets for the average supply of beef, pork, goat and poultry (FAO 2012), a comparison revealing the magnitude of bush meat supply in some African nations. Data on wild meat's relative importance are revealing. During the late 1970s, de Vos (1977) found that in many parts of Africa wild meat supplied at least 20% of all animal protein. Two decades later, Njiforti (1996) reported that in the northern Cameroon annual wild meat consumption of almost 6 kg of bush meat per capita provided nearly 25% of all animal protein intake.

More recent indications are hunting in Western and Central Africa has become strongly market-oriented and many villagers join professional hunters in order to supplement their limited income. According to Kümpel et al. (2010), 60% of poor- to middle-income households in rural

Equatorial Guinea were hunting wild meat for urban markets. Bifarin et al. (2008) found that an average hunter in Nigeria's Ondo state killed every month more than 140 kg of large herbivores (buffalo, bush bucks and duikers) and nearly 90 kg of smaller species (mainly cane and African giant rats). Much, if not most, of this meat is obtained by illegal hunting that is common even in such a relatively well-policed national park as the Serengeti and its adjacent lands where inhabitants of some villages consume annually as much as 36 kg of wild meat per capita, or nearly as much as Japan's average meat consumption (Nyahongo et al. 2005; Holmern et al. 2007; Ndibalema and Songorwa 2008). Large antelopes (eland and topi) are the favorite target but more abundant buffaloes and wildebeest are killed most often. Fa and Brown (2009) concluded that both ungulates and rodents are now hunted at unprecedented rates throughout most of tropical Africa.

The situation is similar in China thanks to the combination of the traditional preferences for eating exotic species and rising incomes that enable millions of consumers to indulge these tastes, often in the form of ostentatious banqueting. Moreover, in 1999, a survey found that 26% of all wild animal dishes served in restaurants contained species on China's endangered list, a practice that led the China Wildlife Conservation Association to ask all professional chefs to sign a declaration affirming that they will not use such animals in their meals (Smil 2004). Endangered species killed for direct consumption or for stocking breeding operations range from frogs, snakes and turtles to birds of prey, ungulates and large carnivores. The extent of this domestic slaughter has led to large-scale illegal imports of wild species from abroad, particularly from Southeast Asia, and to massive commercial breeding; for example, Shi et al. (2008) found that half of China's commercial turtle farms that responded to a survey were selling annually more than 300 million turtles, including some critically endangered species.

Traditional Societies: Animals, Diets and Limits

Traditional (pre-industrial) societies that followed the long period of foraging existence could be assigned to one of three distinct types: pastoralists who did not cultivate any crops but followed their herds of domesticated animals on their migrations to new pastures; shifting farmers who cleared patches of forested land for short periods of crop cultivation before repeating the clearance/cultivation cycle on nearby land; and permanent, sedentary, agricultural societies engaged in progressively more intensive cropping whose surplus allowed a minority of the population to live in towns and cities and, eventually, to prepare conditions for the rise of modern industrial and service economies. The great epochal shift away from foraging was not, as it is still too often asserted, accomplished during a Neolithic "agricultural revolution" originating in one place and then diffusing around the world.

Permanent cropping and the associated domestication of animals had evolved separately in several places in the Old and New World, and gradual intensifications of sedentary agriculture had followed a number of distinct practices and preferences. But all of these arrangements have one key attribute in common. Even those traditional agricultures that were able to produce relatively high yields and that brought the intensity of cultivation close to the maxima supportable by the recycling of nutrients and by adequate irrigation had experienced a high incidence of malnutrition and could not escape recurrence of major famines. These adversities endured even into the early modern era, with malnutrition common in Europe of the 18th century and with major famines recorded not only in India and China but also in Japan during the 19th century (Smil and Kobayashi 2012).

That was the price paid for supporting high, and historically unprecedented, population densities – and the accompanying constant was that an overwhelming majority of people in traditional agricultural societies had to subsist on diets that included only very little meat and that significant minorities led essentially meatless lives. Marginalization of wild meat was an inevitable outcome of agricultural expansion based on large-scale deforestation and conversion of grasslands and wetlands to arable land: seasonal shooting and occasional snaring of small game prorated to only small annual consumption rates.

Large animals – cattle, water buffaloes, yaks, horses and camels – were not domesticated to be primarily sources of meat but to be essential prime movers providing power for field work, food processing, road and off-road transport, and warfare. And the numbers and meat yields of smaller animals, ranging from rabbits to pigs and from pigeons to turkeys, were limited by the quality and availability of feed. Low crop yields and frequent crop failures left little room for cultivating feed crops as even food crop cultivation was dominated by a few grain, legume and tuber staples whose consumption often supplied no less than 80–85% of all food energy.

Moreover, in many societies meat – all kinds of it, or some specific varieties – became a proscribed food. Injunctions against its eating ranged from permanent directives issued by the rulers to an entire nation to elaborate rules specifying the details of abstinence and permissible conduct.

But these bans, some lasting for more than a millennium, could not abolish widespread desire to eat meat and could not topple meat's ranking as a high-status food whose liberal consumption (in forms ranging from deliberate gorging to elaborately prepared meals) was one of the unmistakable privileges of rulers as well as affluent land owners and, later, wealthy town dwellers.

With little progress in average crop yields during the millennium spanning the late antiquity and high Middle Ages, typical diets of peasants (i.e., the majority of pre-modern populations) and urban poor had seen little change as the low rates of meat consumption were dictated by the combination of population densities and environmental and agronomic imperatives, with the limited supply of nitrogen as the most important restriction in the last category of limitations. Some dairy cultures supported modest but widespread consumption of dairy products, but only the richest strata of traditional societies could enjoy meaty diets. Dominant diets were almost meatless even in the industrializing Europe of the early 19th century, and overwhelmingly vegetarian diets prevailed in rural China and India into the 1980s.

Domestication of animals

Domestication is an arrangement where controlled breeding leads to notable changes in the appearance, functioning and productivity of organisms. Domestication of plants, particularly of cereal grains and legumes, secured the largest source of food supply for all subsequent traditional and modern populations, but domestication of animals was not a comparatively less important affair as it provided not only large (in aggregate, if not in per capita, terms) amounts of high-quality food supply, especially as protein in meat, milk and eggs and associated lipids and micronutrients, but also relatively powerful prime movers and sources of manure and raw materials. Moreover, when one thinks about animal husbandry in terms of deferred harvests of high-quality foodstuffs to provide a buffer against crop failures, it is obvious that its management required plenty of strategizing, planning, cooperation and problem-solving, and deployment of these qualities has certainly advanced human evolution.

As is the case with the domestication of plants, we now have a great deal of reliable evidence about the origins of the process – where and when – but continue to speculate about its causes (Digard 1990; Clutton-Brock 1999; Zohary and Hopf 2000; Armelagos and Harper 2005; Driscoll et al. 2009). The two most often invoked causes (triggers) of sedentism and of the domestication of crops and animals are population pressures creating food "crises" (Smil 2008). The case for environmentally driven change (drier and warmer Neolithic climate with higher CO_2 levels) has been made repeatedly, perhaps most forcefully by Richerson et al. (2001): they concluded that agriculture was impossible during the Paleolithic but mandatory in the Neolithic. Given the long coexistence of people and animals, it is remarkable that no domestications took place before about 12,000 years ago – but then the process for the four most important meat species was accomplished just in a matter of centuries.

Goats and sheep were domesticated first, about 11,000 years ago, followed by pigs (10,500 years ago) and cattle (10,000 years ago), with all of these truly revolutionary changes taking place in a crescent-shaped region embracing southern and eastern parts of today's Turkey, northern and northeastern Iraq, and northwestern Iran (Troy et al. 2001; Zeder 2008). Horses were an important source of food for both late Paleolithic and Neolithic hunters, but their domestication – on the Pontic and Central Asian steppes as recently as 2,500 BCE (Jansen et al. 2002) – was primarily for riding and transportation and only much later they replaced oxen as draft animals. Humans have domesticated just a tiny share of terrestrial mammals to be used as sources of food or, with the largest ones, for draft and transport. To the five species just listed that were eventually introduced to every continent, we have to add less than a dozen domesticated mammals with more limited, or only regional, distribution.

Water buffalo was likely domesticated both in India and China about 4,000 BCE; donkeys were tamed first in Egypt (5,000 BCE); dromedary and Bactrian camels came from, respectively, Arabia and eastern Iran, the first one about 6,000 BCE; yaks from the Tibetan Plateau (about 5,000 BCE); and llamas (3,500 BCE) and alpacas are descendants of wild guanaco from the Andes. All of these species were eaten but none of them were domesticated primarily for meat, with transport, draft power, dung, milk and fleece being more important; dogs should be, of course, the most notable entry on this domestication list. To complete the list of domesticated mammals with limited distribution, I must add rabbits (a relatively recent European favorite domesticated only about 1,500 years ago) and guinea pigs domesticated around 5,000 BCE in the Andes.

Why these 15 mammalian meat-producing species and not others? In order to be domesticated, mammals have to satisfy several criteria. They must be able to reproduce in captivity, must be amenable to taming (a prerequisite of domestication), should not be aggressive (intra- or interspecifically) and, obviously, social species with dominant individuals will be easier to coral, herd and breed (Russell 2002). Meat yields and feeding requirements are obvious constraints on animal size and kind. An ideal animal would combine substantial body mass with rapid growth rate (high productivity) achieved by metabolizing widely available, inexpensive feed. In practice, that is an impossible combination. Feed limitations eliminate all carnivorous species; size and feed limitations eliminate extreme sizes, animals that are too small (growing fast but metabolizing rapidly and requiring constant feeding) or too large (maturing too slowly and requiring large amount of feed per head: their specific metabolic rate is low but their mass negates that advantage).

That is why the smallest domesticated meat-producing mammals (guinea pigs and rabbits) have adult mass of at least 1 and 2 kg. Fattening rabbits in cages was a common practice in many traditional European agricultures while guinea pigs have been kept in the Andean region of South America (in Cuzco's cathedral a large painting of the last supper shows Jesus sharing a roasted guinea pig with the apostles) as well as in West Africa. On the other side of the spectrum, slaughter weights of the largest domesticated species (cattle and water buffaloes, both used also for draft and for milking) were no more than about 300 kg (their higher modern weights result from optimized feeding, including non-cellulosic phytomass).

Ruminants (cattle, water buffaloes, yaks, camels, alpacas, goats, sheep) have an inherent advantage as they are able to digest copiously available cellulosic tissues that cannot be metabolized by other mammals and to produce high-quality, protein-rich milk and meat (and blood), foodstuffs that make it possible for the pastoralists to survive in environments that are too arid for cropping and for sedentary farmers to enrich their diets. Perhaps the most generalized definition of pastoralism is as a form of prey conservation with deferred harvests profitably repaid by growing stocks of animals, a trade-off that is more rewarding with larger livestock (Alvard and Kuznar 2001).

Omnivorous pig, the most important animal fed solely for meat and common throughout Eurasia (excepting India), has yet another advantage: its basal metabolic rate is slower than that of sheep, goats or cattle, and hence it converts feed into muscle and fat with higher efficiency. Body mass range of domesticated birds is much narrower as most wild species are simply too small to make rewarding domesticates. Extreme body weights of wild birds that were domesticated for meat (and egg production) range from less than 0.5 kg for pigeons to about 10 kg for wild turkeys (again, modern breeds are much heavier due to breeding and optimal feeding, with maximum male turkey weight in excess of 20 kg). Wild fowl from the tropical forests of Southeast Asia was domesticated about 8,000 years ago, and its descendants are many varieties of chicken, the most numerous domesticated species in the world.

Ducks were domesticated first in China 6,000 years ago. Geese, first domesticated most likely in Egypt about 3,000 BCE, are omnivores that prefer to feed on such succulent forages as clover, blue, orchard and timothy grass but their forced feeding to produce *foie gras* is done with grain, and geese flocks used to be herded to harvested fields to scavenge lost grain, a practice still common in parts of Asia. Turkeys, the only New World avian domesticates, were first domesticated in Mesoamerica some 7,000 years ago, and the breeding of meat pigeons has been common in the Middle East, North Africa and Mediterranean Europe.

But domestication of mammalian and avian species had translated into better diets only for a minority of people. In some cases – as shown by studies of pigs in the highlands of New Guinea and cattle among many African pastoralist tribes (Rappaport 1968; Galaty and Salzman 1981) – animal ownership had to do more with social status (wealth and prestige) rather than with having additional reserves of food. But in most traditional agricultural societies, domestication of animals had resulted in considerably lower meat consumption than was the norm among foragers: higher population densities, limited availability of feedstuffs, and the necessity to use animals for draft and as producers of milk, eggs or wool limited the amount of meat. This reality was one of the major consequences of the great transformation from subsistence dependent on hunting and gathering to food provision dominated by permanent cropping, an epochal shift that led to larger permanent settlements and rising population growth at the expense of average quality of traditional diets.

Population densities and environmental imperatives

That the rise (and subsequent intensification) of agriculture was primarily a response to higher demand for food by growing populations is best illustrated by comparing the carrying capacities of all important means of human subsistence. The lowest densities of foraging populations (in some tropical rain forests as well as in boreal regions) were on the order of 1 person/100 km² of exploited territories; in seasonally dry tropics with the abundance of roots, nuts and megaherbivores, the rates were up to an order of magnitude higher (10/100 km²) and only the groups subsisting on coastal fishing and killing of marine mammals had densities approaching 1 person/km² or 0.01 forager/ha (Smil 2008). Population densities of pastoral population spanned a fairly narrow range of 1–2 people/km² (1–2/100 ha) of grassland, but even shifting farmers cultivating low-yielding crops could feed 10–20 people/km² (0.1–0.2 people/ha), an order of magnitude more than pastoralists and two orders of magnitude more than typical foragers.

And the traditional farmers boosted that performance by another order of magnitude, feeding at least 100–200 people/km² (1–2 people/ha) of cultivated land – and those in warmer climates perfecting the practices of crop rotation, heavy manuring and irrigation could feed three to four times as many people: by the end of the 18th century, South China's rice-growing region averaged 5.5 people/ha (Smil 2008). The peak performance of such agroecosystems was impeccably documented by Buck's (1930, 1937) surveys of traditional Chinese farms during the 1920s and the early 1930s when large parts of South China could support 7 people/ha and when the most intensive double-cropping was able to feed more than 10 people/ha.

But it must be stressed that all of these high population densities were made possible not only because of intensive cropping but also because of virtually meatless diets despite common presence of numerous domesticated mammals and birds. Three long-lasting realities explain this apparent contradiction: the need for large animals as the only relatively powerful mobile prime movers; the combined value of resources steadily provided by both large and small animals (milk, manure, fleece) being greater than the one-time provision of meat; and a low productivity of slowly growing, small-stature domesticated breeds as the limited crop yields left no room to divert scarce cereals, legumes or tubers from human food to feeding those animals kept only for meat (pigs above all).

Traditional society had no mobile inanimate prime movers on land: its only two mechanical devices able to convert water and wind flows – water wheels (whose slow diffusion began in antiquity) and windmills (whose widespread adoption dates only to the Middle Ages) – were stationary, heavy and relatively inefficient (Smil 2008). And yet traditional societies had many demands that could not be met, effectively and efficiently, only by deploying human muscles: these tasks ranged from pulling out stumps in lands converted from forests to cropland to building large cathedrals and from plowing of heavy clay soils to long-distance transport of heavy burdens.

These, and many other, tasks were made possible by domesticating large animals for transport (pulling heavy wooden carts) and for the performance of field labor (above all plowing and harrowing), food processing (threshing, grain and oilseed milling) and other mechanical tasks aiding the local extractive or manufacturing activities (water pumping, crushing or grinding by animals walking in circles harnessed to whims). Slower but less demanding ruminants were cheaper to use for these tasks than horses (oxen could survive only on roughages while working horses required at least some grain), and hence oxen remained the most important draft animals even in richer parts of Europe until the late Middle Ages, with horses being used primarily for riding (including messenger services) and warfare. Large, and slowly reproducing, draft animals were thus too valuable to be slaughtered for meat once they reached their maximum weight.

Moreover, in the absence of any synthetic inorganic fertilizers, all animals were a voluminous source of relatively concentrated nitrogenous compounds. While other organic matter that was recycled in order to boost nitrogen content of farmed soils had only about 0.5% (cereal straws, legumes plowed under as green manures) or between 1 and 1.5% (legume straws, human wastes) of the nutrient, animal manures contained commonly more than 2% and some, especially pig manure and poultry waste, even more than 3% of nitrogen. Animals as steady source of fertilizer were of greater value than animals killed as soon as possible as one-time sources of meat and that is why all milk-producing animals were kept as long as possible in order to provide both soil nutrients and high-quality protein in dairy products.

Finally, productivity of animals – those killed at the end of their useful lives as draft or transport beasts or as producers of milk, as well as the minority kept only to produce meat – was low because their feed was generally poor and because they had to spend a great deal of energy in searching for it and gaining weight slowly. Pigs, omnivores *par excellence*, were fed any available food processing, farmyard or household waste. Kitchen, and in more recent centuries also restaurant, waste has been an important source of pig feed for small-scale producers, including those in urban and peri-urban operations, with no example of this practice being more remarkable than that of Coptic *zabbaleen* of Cairo (Maike 2010). But they lost a major source of income in 2009 when the government ordered slaughter of the city's pigs, ostensibly to prevent any spread of swine influenza (H1N1) although the animals were not infected.

Free-roaming pigs or animals fed waste scraps took well over a year to reach their (lower than today's) slaughter weight, compared to less than six months for today's confined animals fed optimal carbohydrate and protein mixtures. Similarly, chicken and ducks were mostly left to themselves to feed on insects, soil invertebrates, plants and wild seeds, with occasional supplement of grain, and in some regions geese were herded to forage in harvested fields. Free-roaming chickens were killed at three or four months of age (instead after just six weeks today). Such meat was not only tougher but necessarily more expensive, and its frequent consumption was out of reach of most peasants, many of whom had limited access even to cheaper milk and cheeses. There is plenty of archaeological and written evidence to confirm the long duration of these low-meat diets in all traditional societies.

Long stagnation of typical meat intakes

As I have already noted when citing the evidence from Çatalhöyük, some early agricultural societies continued to consume substantial amount of wild meat, and stable isotope evidence shows that the inhabitants of this Neolithic settlement also ate domesticated goats and sheep and consumed their milk (Richards et al. 2003). Archaeological evidence from the middle Rhine region shows that the Neolithic populations practiced extensive grain cultivation – but Dürrwächter et al. (2006) reported that most individuals had $\delta^{15}N$ values consistent with intakes of significant amount of animal protein. Isotope analysis by Oelze et al. (2011) found that early farmers at another Neolithic site in Central Germany ate an omnivorous diet of C_3 plants and animal meat from livestock. Similarly, proteins from terrestrial, rather than from aquatic, animals dominated the diet of Neolithic populations in southern France (Herrscher and Le Bras-Goude 2010), and even a Greek colonial population on the Black Sea coast of Bulgaria derived much of its protein from terrestrial sources (Keenleyside et al. 2006).

European historical record is clear: although the typical meat intakes were low, meat was a highly regarded food whose consumption played important roles in social, religious and military affairs (Swatland 2010). Ancient Greek texts leave us with the impression of mythical meat-eating heroes, but both Greek and Roman diets were dominated by grains and legumes, and Purcell (2003) noted that in Rome meat was seen as a prime product of the sacrificial economy, but cured meats (ham, sausages and bacon: *perna, farcimina* and *lardum*) were essential foods of the Roman army (Roth 1999). But overall consumption was low: when Allen (2007) constructed two different consumption baskets representing annual requirements of a respectable and a poor working Roman family, he assumed that the former one consumed 26 kg of meat per capita while the latter ate only 5 kg/capita, most likely a generous assumption. And as the population densities slowly increased, the quality of average diets had declined.

This process cannot be quantified in terms of actual food intakes or shares of meat consumption, but its consequences have been well documented in archaeological record. Lower meat intakes had almost always translated into the lower availability of high-quality protein (in some cases, it was compensated by eating more fish and dairy products) as well as into the lower supply of several vitamins (A, B12, D) and minerals (above all iron). These shifts in diet quality were reflected in diminished statures of sedentary populations (Cohen 2000; Kiple 2000). Koepke and Baten (2005) measured skeletons of 9,477 individuals who lived in Central and Western Europe between the 1st and the 18th centuries as a proxy of the biological standard of living and found that neither male nor female heights increased during the entire period of the Roman Empire and that the Roman migrants from the Mediterranean region to Central Europe were, on the average, 4 cm shorter than the local populations.

Lower intakes of high-quality animal proteins are the best explanation of these differences as the seasonally arid Mediterranean climate precluded any extensive cattle grazing and low crop yields prevented any significant diversion of grains or legumes to animal feeding. Excavated animal bones tell a story of declining beef consumption in Rome, with cattle share falling from 28% during the 1st century BCE to about 8% between the 1st and the 2nd centuries CE and to nothing in the 2nd and 3rd centuries, and a substantial rebound (and significant height increase of people) only with the 5th century Germanic invasions. But, as Koepke and Baten (2005, 61) point out, the overall record is one of stagnation and "there was no large-scale progress in European nutritional status …, not even for the period between 1000 and 1800, for which recent GDP per capita estimates indicate increasing development."

Similarly, Fogel (2004) stressed how "technophysio evolution," through which humans gained an unprecedented degree of control over their environment, enabled more than 50% increase in average body size (and doubling of average longevity) since 1800, with most of the gains taking place only since the closing decades of the 19th century. He also showed that conquest of high mortality and hunger in Western Europe did not begin until the 1780s and that in England and France it was over, respectively, only by the 1830s and the 1840s. This does not mean that during certain periods segments of the population in some parts of the continent did not eat highly meaty diets.

Fragmentary medieval documentation is less amenable to generalization, but abundance of testimonies, observations and lists from the 16th and 17th centuries makes it clear that all rich households in the early modern England and Scotland had excessively meaty diets, that the continental visitors were generally astonished at the quantities of served meat and that plenty of meat was eaten also by well-off professionals including lawyers, doctors or master craftsmen (Rixson 2010). Quantities consumed can be perhaps best appreciated by referring not to an annual supply of the royal court (1,240 oxen, 8,200 sheep, 2,330 deer, etc. at the time of

Henry VIII) or large aristocratic holdings but to meals described by the period's best-known diarist, Samuel Pepys.

On April 4, 1662, he and his wife (and their servants) had "a fricassee of rabbits and chickens, a leg of mutton boiled, three carps in a dish, a great dish of side of lamb, a dish of roast pigeons, a dish of four lobsters, three tarts, a lamprey pie ... all things mighty noble, and to my great content." But Pepys was hardly an ordinary man (being a high Navy administrator moving in the company of the country's ruling elite), and as Thirsk (2006, 235–236) notes, those envious continental visitors "did not form their judgments from encounters in the deep countryside or by entering homes of ordinary folk." If they would have looked in the pots of ordinary husbandmen, the meat they contained was most likely sheep's head, calves' feet, meaty bones or a slice of smoked fat bacon.

Quantifying this long-lasting stagnation in terms of actual average per capita meat intake is impossible. While archaeological searches and written documents offer a wealth of information about the composition of diets in the antiquity, Middle Ages and the early modern era (Thirsk 2006; Wilkins and Hill 2006; Dawson 2009), this evidence is mostly anecdotal or fragmentary or it is restricted to specific locales and hence it precludes converting such numbers or trend into revealing summaries of broader regional or national developments. But we know that armies consumed plenty of meat (a Tudor soldier was provisioned with 2 lb of beef or mutton a day) and even some prisoners were not badly off: in 1588, the Bury House of Correction had a daily allowance of 0.25 lb of meat or about 40 kg/year. Another major uncertainty precluding any reliable quantification concerns the typical body weights: for example, during the 16th century, English beef cattle may have averaged only about 150–160 kg and mature pigs only 40 kg, in both cases less than half of today's slaughter weights.

In any case, British diets changed for the worse by the end of the 18th century: the country's population was more than 60% higher than in 1650 and a series of Enclosure Acts transferred large areas of common land to private estates, leading to abandonment of many villages, destitution of peasants and poor food supply conditions resembling those on the continent (Spencer 2000). This is easily confirmed by economic realities, personal observations and actual consumption estimates and calculations from the late 18th and the early 19th centuries when meat was still a rare treat not only in ordinary English households but also in Europe's largest continental nations, in France and Germany.

During the last decades of *ancien regime*, France, the greatest continental power, had low per capita level of food production (reflected in small statures and low work capacities), high mortality (above 35/1,000 by 1750), chronic malnutrition and monotonous diets. Antoine Lavoisier (1791) noted in his treatise on the riches of France that large numbers of peasants ate meat only at Easter or when invited to a wedding. The best reconstruction of average French food intakes at the beginning of the 19th century indicated that meat consumption contributed less than 3% of all food energy (Toutain 1971). James Paul Cobbett, traveling through France in 1824, repeatedly commented on how much more advanced was the British farming and rural life, noting that "a less quantity of meat is requisite to a French labourer than what labourers (when they can get it) are used to consume in England. The economy in cooking here is such that the same quantity of animal food which we eat in England would feel almost double the number of persons in France" (Cobbett 1824, 189).

That is a telling comparison considering that when Sir Frederick Morton Eden surveyed the state of England's poor, he found that even in the richer southern part of the country they "are habituated to the unvarying meal of dry bread and cheese" and "if a labourer is rich enough to afford himself meat once a week, he commonly roasts it" (Eden 1797, 100). A laborer in Leicestershire (in the Midlands) told him that in his household they had "seldom any butter, but occasionally a little cheese and sometimes meat on Sunday.... Bread, however, is the chief support of the family, but at present they do not have enough, and his children are almost naked and half starved" (Eden 1797, 227). Eden's report was also used in a reconstruction of per capita food intakes by poor English and Welsh rural laborers between 1787 and 1796 that came up with an average of 8.3 kg/year (Clark et al. 1995), and even during the 1860s meat consumption of the poorer half of the English population was barely above 10 kg (Fogel 1991). According to Abel (1962), average German per capita meat consumption was less than 20 kg before 1820.

And even lower consumption rates prevailed in Asia's populous nations well into the 20th century. While Japan was an exception, meat was highly appreciated in both traditional China and Korea (Nam et al. 2010) and it was eaten as much as it could be afforded, which in most cases amounted to very little. Reliable data gathered by Buck's (1930) exhaustive study of agricultural practices and food supply in all of China's major provinces during the 1920s make it clear that meat remained a rarity in peasant diets well into the 20th century. In Hebei province, annual meat intakes were as low as 1.7 kg/family or less than 300 g/year/capita. This means in China's poor arid northern provinces, most peasants (much like their 18th-century French counterparts) ate only a few morsels of meat two or three times a year, usually at the New Year's celebration and weddings.

Meat consumption was substantially higher in the richest provinces, reaching the peak at just over 30 kg/family (or about 5 kg/capita) in Jiangsu. Buck's later (1929–1933) surveys in 22 provinces indicated that working adult males averaged about 80 kcal of meat a day, or about 8 kg of meat and lard per year, and the mean for the entire population was below 3 kg/capita (Buck 1937). During the early 1930s, meat supplied only about 2% of China's average per capita food energy intake (similar to the French rate of the 1780s), or only about half as much as white potatoes (disliked in China) and less than a third as much as legumes (mostly soybeans), a major source of protein.

Limited supply of meat in traditional societies also led to two very different ways of treating the food. First, in many cultures meat consumption was forbidden for religious reasons, completely and permanently (as in the early Buddhist Japan) or selectively (applied to beef or pork) and temporarily (during specified periods or days of the Catholic calendar). Meat would have been out of the reach for the majority of people in any case, but a religious ban helped to elevate those inescapable, and unwelcome, shortages to the category of laudable and morally superior sacrifices. Second, as is usually the case with a scarce commodity, meat consumption became a sign of prestige and privilege. I will briefly address both of these phenomena before commencing Chapter 3 that deals with the transitions from traditional to modern diets and with meat production and consumption in modern societies.

Avoidances, taboos and proscriptions

Even traditional hunters who could not be that choosy about the sources of their food energy were selective and had never used all meat sources that could be collected, trapped or hunted in their environment. Animals were always avoided not only because they were inedible but because of reasons ranging from poor energy returns on risky pursuits (hunting leopards or cheetahs or even larger and more aggressive carnivores) to the veneration or abhorrence of certain species considered sacred, unclean or disgusting. As already explained, body mass, productivity and feeding requirements have limited the number of domesticated animals to only a tiny share (on the order of 1% for mammals and only 0.1% for birds) of vertebrate species. But what is so remarkable about meat consumption in settled societies is that in so many of them the avoidance extended to species whose eating was relished by other groups and cultures.

This culturally driven avoidance had evolved along two major lines, as informal taboos (unwritten social rules, often of ancient origin and remarkable durability) and as formal proscriptions elaborated by dominant religions and often specified in many (irrational but dutifully followed) details. Extensive research on eating, food taboos and religiously motivated bans offers many speculations and explanations about their motives and causes and about the factors that facilitate their acceptance and long-term adherence to what must have been often onerous restrictions. Grivetti (2000) surveyed hypotheses explaining the origins of these dietary codes that included esthetic and compassionate considerations, divine commandments, and other spiritual and magical and religious impulses as well as more practical motives of resource conservation or an effective management of limited food supply.

Specific taboos – be they applicable to all people of a particular culture at all times, or only to some of its members (children, adolescents, women) or some of them at special times (menstruating or pregnant women) – can be found in every culture and affect foods in every major category, from fruits to meat. These specific meat taboos have often denied high-quality protein to women and even to children (Meyer-Rochow 2009). Chinese taboos concerning diets in pregnancy are particularly complex (Lee et al. 2009). Some of these ancient rules forbid, among many other foodstuffs, consumption of beef, mutton, deer, chicken and pheasant.

The best-known religiously motivated examples of foodstuffs are the avoidance of pork in Judaism and Islam, of beef in Hinduism and of all meats in the strictest interpretation of Buddhism. Origins and evolution of these proscriptions have been widely debated by anthropologists and historians (a few notable contributions are Harris 1966, 1974; Diener and Robkin 1978; Simoons 1979, 1994). Judaic restrictions – eating only ruminants (chewing the cud) with cloven hoof – clearly reflect the biases of a pastoral society in arid Middle Eastern environment: cows, goats, sheep, deer, bison, gazelle, antelope, ibex and addax, meats eaten by pre-Israelite pastoralists, are in (so are giraffes!), but hares, camels, pigs, donkeys and horses are out (*beheimah temeiah*, unclean), as are all carrion, mortally injured animals, all blood, and fats on internal organs and sciatic nerves of *kashrut* animals (Garfunkel 2004). Similarly, Muslim rules forbid pork, blood and carrion, and both religions insist on ritual slaughter (*halal*). Besides beef, Hindu *Caraka Saṃhitā* set down about 1,500 BCE, forbids eating young dove, frogs and all carrion (Valiathan 2003).

Complete bans on eating all meat, permanently or temporarily, are more surprising and are harder to explain. They had evolved only in Buddhist and Christian societies, but different countries within those two religious realms had adhered to them with varying, and changing, vigor. In no other complex traditional society was meat consumption affected more by religiously driven proscriptions than in Japan. Limited grazing land on the archipelago's three main islands (Hokkaidō with its extensive grazing lands was not integrated into the empire until the 19th century) and the necessity to produce staple grain crops (rice and barley) in alluvial lowlands were two natural factors that would have restricted the availability of animal feed in any case, but the main reason for minimal meat intakes were the repeated bans of meat eating rooted in the Buddhist faith. *Ahimsa*, the Buddhist avoidance of violence, is defined by Vyasa's commentary on *Yoga Sutras* as "the absence of injuriousness (*anabhidroha*) toward all living things (*sarvabhuta*) in all respects (*sarvatha*) and for all times (*sarvada*)," and it is a core concept of the faith (Chapple 1993).

The first imperial prohibition of meat eating came in 675 CE, more than a century after Buddhism reached the islands from China via Korea, and repeated proclamations followed during the next century and during the Kamakura period during the 14th century (Ishige 2001; Watanabe 2008). In addition, there were also Shintō taboos on the eating of cattle, horses and fowl. Although some wild animals were always consumed in mountainous areas, more widespread meat eating commenced only with the spread of Jesuit missions during the second half of the 16th century, but the bans on killing and eating animals came back during the early decades of the Tokugawa shogunate (1603–1867). The last decades of Tokugawa rule saw growing sales of wild meat (boars, deer, hares) in specialty stores in large cities, but its consumption was still mostly as a medicine for the sick. Only the restoration of the imperial rule and a rapid opening to the West reintroduced the country to slowly rising meat production and consumption (Krämer 2008).

In 1871, the government declared meat consumption to be important for good health. A year later, an official announcement disclosed that the Emperor consumed meat regularly; moreover, meat became not only a part of army rations but even Buddhist monks were allowed to eat it (Ishige 2001; Watanabe 2008). But for decades meat eating remained a largely urban phenomenon. In 1900, the nationwide average of annual per capita meat supply (in terms of carcass weight) reached only about 800 g/capita (mostly beef), which means that most people still never ate meat; by 1925, the mean rose to about 1.7 kg and in 1939 it was 2 kg/capita, still a negligible source of protein and lipids.

Early Christians, reacting to Judaic proscriptions, saw all foods as edible but blood, carrion and meat killed as part of pagan idolatry was soon forbidden. Asceticism of many early desert Christian communities, drawing inspiration from St. John's diet of nothing but *akrides* and *meli* (meaning of both remains obscure, locusts and honey being a misrepresentation). A less restrictive regimen was formulated for the monks by

Saint Benedict of Nursia (480–547 CE), and *Sancti Benedicti Regula* became the dominant guide of monastic living. According to the ninth part of the 39th chapter (*De mensura ciborum*) "*Carnium vero quadrupedum omnimodo ab omnibus abstineatur comestio, praeter omnino debiles aegrotos*" (Except the sick who are very weak, let all abstain entirely from eating the flesh of four-footed animals).

During the Middle Ages, a complicated set of rules governed the periods of fasting, with abstinence from red meat or dairy products during Lent and the Advent, and with meatless Wednesdays, Fridays and Saturdays (Grumett and Muers 2010). Consequently, red meat was forbidden for more than half of all days in a year and, not surprisingly, with time those who could afford to break the rules found ways to wean or to circumvent the rules, and this led eventually to their considerable relaxation. As a result, by the end of the Middle Ages most well-off religious communities in England were eating a diet very similar to that of well-off secular households (Harvey 1995). Similarly, in Japan a creative interpretation labeled wild boar as mountain whale (*yama kujira*), and hence free to be eaten.

Curiously, one of the strongest, albeit unwritten, meat taboos is the avoidance of eating horse meat in North America. Horse was eaten by Paleolithic and Neolithic hunters, but in those traditional agricultural societies where it was kept for draft power, only old and sick animals were slaughtered and eaten and horse meat was a marginal source of energy and protein. Europe's hippophagic nations are both in the western (France, Belgium, Germany, Switzerland) and eastern (Poland, Ukraine, Romania, Hungary) part of the continent, and elsewhere horses are eaten throughout Latin America, in Central Asia (Kazakhstan, Mongolia) and in Korea and Japan, but in the US even the slaughtering of horses for meat exports has been prohibited (by an act of Congress no less) since 2007.

Meat as a prestige food

On the one hand, not much has to be said in this respect: meat in general, and its specific kinds or cuts in particular, has been seen, by most cultures and by all but a small fraction of people, as a desirable, prestigious food whose copious consumption confers, and is associated with, a high(er) social status.

On the other hand, extravagant practices of meat eating could be a topic for a lengthy book. Meat was a high-status food during the long evolution of our species as is well attested by the fact that for the hunters and the leaders of foraging societies meat sharing, and often minutely specified rules of meat distribution, was an important means of sustaining familial and tribal cohesion and alliances (Patton 2005).

These practices were further elaborated in many pastoral societies where meat was the meal and the status was by meat consumption, and where meat fed "not only the bodies, but also the minds" and was "the very idea of food itself" (Lokuruka 2006, 206). The people who developed the etiquette of prestige meat sharing to its most elaborate level were the cattle-owning tribes of East Africa: for them, wild meat and poultry were marginal sources of food, and sharing the slaughtered livestock was used to maintain good relations, to achieve power and recognition as meat became the currency of social obligations and identity.

Lokuruka (2006) describes the precedence order for meat portions distributed during the traditional, and now much less common, males-only feast of NgTurkana of northern Kenya in detail. The eldest participant got the left femur and attached muscles, the second eldest male received the tibia and muscles, the third eldest male present in the party was awarded the right humerus and attached muscles and the right scapula, the fourth eldest male got the right radius/ulna, carpus, metacarpus, phalanges and their attached muscles – and so on all the way to the seventh eldest man in the party. Organs were also shared in rigid order, with the animal's colon going to the eldest male.

Throughout history, ostentatious, often excessive, consumption of meat was a tangible and tasty demonstration of power and riches as (depending on the cultural setting) entire roasted joints of large animals, servings of more than a score of smaller species presented as entire cooked carcasses (sometimes even reconstituted with their original body coverings) or long processions of elaborately prepared meat dishes were the centerpieces of meals, feasts and banquets. Meat retained its high status in antiquity. In Athens, meat eating by the public was associated with exalted events as it derived from large animals that were sacrificed as a part of communal religious ceremonies (Osborne 2004). In Rome, ceremonial eating of sacrificial meat was restricted to upper classes, and only the lower quality leftovers were sold to the public (Garnsey 1999).

The circle of those with access to plenty of meat changed slowly; since the antiquity, it has always included the rulers. During the Middle Ages, frequent eating of meat (and in interior locations also of marine fish) was expanded to several strata of the nobility, the upper clergy, affluent merchants and rich town dwellers, often in the form of conspicuous consumption in banquets and celebratory feasts (Woolgar 2001). In Rome's case, the entire papal capital enjoyed a superior diet: in 1641, a Vatican manuscript noted that Rome consumed twice as much meat as Naples, a city twice as large, a privilege that did not last as Rome's per capita meat consumption fell by nearly half between 1600 and the 1780s (Revel 1979). Privileged consumption of wild meat by the rulers and by nobility was reinforced by hunting bans and severe penalties for poaching, and high cost of domestic meat (with the exception of low-status pork) denied all but an occasional consumption to all poorer strata of medieval societies.

But for those who could afford them, medieval meat choices were much wider than today. Beef and pork dominated, but in many regions mutton and goat were common, and the consumption of avian species was not restricted to domesticated poultry. One of the most revealing list of meat-eating excess enumerates the quantities of meats consumed during the banquet honoring the installation of George Neuvile as Archbishop of York under King Edward IV in 1466 (Flandrin 1989). Besides 104 oxen, 6 wild bulls, 1,000 sheep, 304 hogs, 2,000 suckling pigs and large numbers of chicken, geese, ducks and pheasants, 2,500 invited guests were also offered 400 swans, 400 plovers, 104 peacocks, 204 cranes, 400 herons and 1,000 egrets. Dawson's (2009) detailed examination of English habits showed that little had changed during the 16th century when beef and mutton dominated meat supply, venison was a favorite, and birds trapped and shot for eating ranged from ordinary sparrows and larks to a wide range of aquatic species including herons, cranes and curlews.

Hierarchies of high-status meat, both as far as the species and the cuts were concerned, had shifted over time. During the Middle Ages, game, aquatic fowl and marine mammals were the high-status meats while butchered meat went into stocks, minces and stews. By the middle of the 17th century, whale, dolphin and seal meat disappeared from feasts, as did swans, egrets and herons, and by the 18th century there was a clear distinction between *basse boucherie* for *basse peuple* (such as pig's snout or belly) and the fancy cuts for fine dining (barely cooked steaks). Along the way, meat dishes highly flavored with imported spices (in peak favor during the 14th and 15th centuries) gave way to meats prepared with native herbs. In the early modern Italy, milk-fed veal (*vitella mongana*) was at the top, costing 2.5–3 times as other meats (Revel 1979). In the early 17th-century France, veal was no more expensive than mutton and pork, and all of these three meats were more expensive than beef – but then came, rather swiftly, *la réhabilitation du boeuf* (Flandrin and Montanari 1996).

The great dietary transition that will be described in Chapter 3 has increased the access to meat for hundreds of millions of consumers, but it did not eliminate meat's high dietary status. In a new setting where previously expensive meat varieties became widely affordable, it is not, as it was in the past, the quantity of overall meat intake that is the distinguishing variable but the kind of meat that is consumed (with chicken experiencing the most remarkable global ascent) and the quality of meat cuts, a matter of some fundamental disagreements, with highly fatty beef retaining its superior status in Japan and lean cuts of all meats becoming the top choices of Western consumers of higher socioeconomic status (Darmon and Drewnowski 2008).

Japan is perhaps the best example of dietary transition leading to a relatively high meat consumption in a society where meat eating was actually banned for centuries. The country's most prized meat, highly marbled Matsuzaka beef, costs more than $200/kg. Photo by V. Smil.

3

Meat in Modern Societies

In Europe, North America, Australia and Japan, the twinned processes of industrialization and urbanization accelerated during the closing decades of the 19th century; elsewhere they began to unfold rapidly only several generations later – in India and China, the world's two most populous nations, only after WW II. The two processes brought a number of fundamental social, economic and environmental changes that, in turn, had many influences on their progress – and dietary (nutritional) transition has been one of the most important, although generally overlooked (particularly when compared to the attention given to economic, technical and political developments), components of this multifaceted creation of the modern world. The transition consists of a necessarily gradual, but in a historical perspective often fairly rapid, shift from traditional, on average barely adequate and monotonous, diets dominated by carbohydrate staples (grains or tubers supplemented by legumes) to food supply well in excess of prevailing metabolic needs able to provide diets of unprecedented variety and quality (but too often containing also too much fat and sugar).

Higher meat intakes were a universal component of the transition, the process driven primarily by rising disposable incomes and in some European countries aided by a rapid expansion of long-distance meat trade. I will look in some detail first at the transition's progress in Western countries, then at its latest phase in rapidly modernizing economies of Asia and Latin America and finally at the globalization of

Should We Eat Meat?: Evolution and Consequences of Modern Carnivory,
First Edition. Vaclav Smil.
© 2013 John Wiley & Sons, Ltd. Published 2013 by John Wiley & Sons, Ltd.

tastes, a process that, once again, has an important meat-consumption component, particularly evident in the popularity of consuming relatively inexpensive meat-based meals prepared by global chains of fast-food outlets.

Afterward I will turn to the modalities of modern meat production and review the principal production arrangements that are still dominated, as far as the total numbers of animals are concerned, by various pastoral practices and by livestock kept as an integral part of mixed farming. But in terms of total meat production, the intensive, "landless" practices where animals are raised to their slaughter weight in confinement already dominate the supply of pork and poultry. This transformation, severing livestock production from its traditional setting in complex agroecosystems and turning it into a mass-scale factory-like enterprise, has been made possible by supply of modern concentrate feeds formulated for optimum growth and meat gain (and imported not only from other parts of a country but often from other continents) and by improved conversion efficiencies of feed into meat.

These arrangements have resulted in poor treatment of animals (and in activism aimed at exposing and reforming those practices) and required new ways of large-scale slaughtering, meat processing, storage, distribution and retail. An increasingly centralized nature of these activities has improved both the quality of meat and the completeness and quality of meat production statistics, but it has not eliminated uncertainties of international comparisons that arise from using non-uniform reporting standards. Similarly, continuing uncertainties affect our understanding of actual meat consumption and associated waste of meat at retail, food service and household levels. I will address all of these concerns before turning to the environmental consequences of massive carnivory in Chapter 5.

Dietary Transition: Modernization of Tastes

Key components of modern dietary transition have been surprisingly uniform around the world: declines and eventual leveling off at much lower rates mark the consumption of traditional carbohydrate staples (cereal grains, tubers) and legumes (consumption of these major sources of protein in traditional diets receded even faster than that of cereals). Increases and eventual establishment of new high plateaus mark the consumption of animal foods in general and meat in particular (meat, fish, eggs, dairy products) and of a greater variety and a better quality of fruits and vegetables. The three most fundamental factors behind the rising meat consumption have been the worldwide adoption of mechanical prime movers in agriculture, availability of inexpensive synthetic nitrogenous fertilizers, and new varieties of crops that doubled, even tripled, traditional yields and released more farm-land previously used to produce food crops for growing high-quality feed crops.

Displacement of draft animals by internal combustion engines (powering tractors, combines and trucks) and electric motors (powering pumps and crop processing machinery) was completed in the West by the 1960s, the decade when the process began in many parts of Asia where it now nears its completion. Synthetic nitrogenous fertilizers became first available before WW I, but their large-scale adoption had to wait until after WW II: their high applications, augmented by phosphatic and potassium compounds, have been critical for achieving and sustaining high crop yields and for greatly reducing the importance of domestic animals as sources of reactive nitrogen. Higher crop yields made it possible to use more land for animal feed: at the beginning of the 20th century, just over 10% of the global grain harvest was fed to animals (most of it to horses, oxen and other draft animals, not to produce meat), by 1950 that share rose to 20% and by the century's end it was above 40% (Smil 2001a).

Other technical advances that promoted higher consumption of meat have included refrigeration and the combination of efficient diesel engines and large ships that ushered in the era of inexpensive seaborne trade. Large-scale commercial refrigeration allowed for unprecedented economies of scale in slaughtering and distribution, while household refrigerators (and also freezers) made meat purchasing and cooking more convenient, less wasteful and less expensive. Emergence of intercontinental meat trade, first as chilled meat, later also as massive shipments of live animals, has been another important factor in making meat consumption more affordable.

Onsets of dietary transitions varied by more than a century, beginning in North America and Western Europe during the latter half of the 19th century, affecting Mediterranean Europe only after 1900 and expanding to the richest countries of East Asia only after 1950, and to many other modernizing nations only during the past generation, with China being the most remarkable case of accelerated change. Specific trajectories from traditional diets of predominantly rural societies to modern diets of largely urban populations have been studied both in high-income post-industrial Western economies and in many rapidly modernizing nations including China, India, Indonesia and Brazil (Bengoa 2001; Caballero and Popkin 2002; Lee 2002; Shetty 2002; Lipoeto et al. 2004; Smil 2004; Weng and Caballero 2007).

The transition has been completed in most affluent nations (where a new dietary transition is under way), but its intensive phase is yet to begin in the poorest countries of sub-Saharan Africa where it is in evidence only among the richest segments of urban populations. Consequences of the dietary transition have been wide-ranging, from obvious improvements in general health of better nourished populations, their increasing longevity and improving quality of life to major transformations of many economic sectors going far beyond crop cultivation and animal husbandry. These industrial and service sectors have included synthetic fertilizers, other agrochemicals (pesticides, herbicides, fungicides), construction of agricultural machinery, road, train shipping transportation of bulk farm commodities as well as refrigerated or frozen foodstuffs.

International and global consequences of the transition have included the emergence and deepening of new trading patterns, and intensified food production and trade have been responsible for environmental changes on scales ranging from local to global. On the individual consumption level, annual meat intakes adding up to more than the body mass of adults (i.e., to more than 60–75 kg/capita) have become a new norm in many affluent countries, a development with many economic, social and health implications. And a rapidly progressing globalization of tastes has diffused new universal (or nearly so) meaty meals ranging from hamburgers and hot dogs to fried chicken and pizza with processed meat toppings.

Urbanization and industrialization

Urbanization and industrialization had served both as the two leading driving factors of higher demand for better nutrition in general and for more animal foods in particular and as enablers of higher agricultural productivity and of other technical and managerial advances that made it possible to produce more meat and to deliver it to new markets. This complex feedback changed the millennia-old composition of labor force in just a few generations. In 1850, three-quarters of America's labor force were still working on the farms, by 1870 that share fell below 50% and by 1900 it was just over one-third, even as the total size of the country's population had more than tripled compared to 1850 (USBC 1975). Similarly, rapid labor transitions marked the progress of urbanization in the richest European countries as the rural labor was released due to the progressing mechanization of field work (relying on machines ranging from mechanical reapers to first grain combines) and food processing (from steam-driven threshing to flour milling).

Wider availability of meat from domesticated animals rested on removing or changing the three great obstacles that prevailed throughout the pre-industrial era: the necessity to breed and maintain large animals as the only prime movers in agriculture and road transport; the necessity to combine cropping and animal husbandry in order to get a steady supply of animal manures, the only relatively concentrated source of nitrogen; and relatively low crop yields that made it impossible to allocate more land for the production of high-quality feeds without compromising the supply of staple grains and legumes. The need for draft animals in farming remained unchanged until the beginning of the 20th century: serial production of tractors began in the US only in 1905, and the number of horses and mules working in US agriculture peaked only in 1919. Subsequent adoption of tractors was rapid, and by the end of the 1920s, their power capacity surpassed that of America's draft horses, releasing large areas of grassland and arable land for grazing of meat or dairy animals and for production of animal feed. In Europe, the rapid adoption of tractors came only after WW II.

The second transformation started with the synthesis of ammonia and the production of inorganic nitrogenous fertilizers just before WW I (Smil 2001a). Here, Europe was ahead of the US, but in both cases widespread and intensifying use of synthetic nitrogenous fertilizers came only after WW II. Manures became a marginal source of the nutrient, and their contribution declined even more as mixed farming gave way to specialized cropping and separate and concentrated meat production. The third change came in two phases, the first one a function of higher crop yields that allowed a rising share of high-quality harvests to be diverted for feeding, and the second one on a large scale only after WW II with confinement of meat-producing animals fed optimized mixtures.

Food demand was stimulated not only by higher disposable incomes of urban dwellers but also by their need for more convenient foodstuffs resulting from a higher participation of women in labor force, less time available for cooking and preference for foods that could be prepared rapidly or that were ready to eat, leading to a great expansion of processed meat industries. Rapid expansion of the late-19th- and the early-20th-century meat processing was done in poor hygienic conditions. These often horrendous realities were graphically attested by Sinclair (1906) in *The Jungle*; an

extended quote describing sausage-making is a must in order to appreciate
the practices at the beginning of the 20th century:

> There was never the least attention paid to what was cut up for sausage;
> there would come all the way back from Europe old sausage that had been
> rejected, and that was moldy and white—it would be dosed with borax and
> glycerine, and dumped into the hoppers, and made over again for home
> consumption. There would be meat that had tumbled out on the floor, in
> the dirt and sawdust, where the workers had tramped and spit uncounted
> billions of consumption germs. There would be meat stored in great piles
> in rooms; and the water from leaky roofs would drip over it, and thousands
> of rats would race about on it. It was too dark in these storage places to
> see well, but a man could run his hand over these piles of meat and sweep
> off handfuls of the dried dung of rats. These rats were nuisances, and the
> packers would put poisoned bread out for them; they would die, and then
> rats, bread, and meat would go into the hoppers together. This is no fairy
> story and no joke; the meat would be shoveled into carts, and the man who
> did the shoveling would not trouble to lift out a rat even when he saw one—
> there were things that went into the sausage in comparison with which a
> poisoned rat was a tidbit. There was no place for the men to wash their
> hands before they ate their dinner, and so they made a practice of washing
> them in the water that was to be ladled into the sausage.

New laws and regulations were passed – Sinclair's book had clearly
helped to adopt the Meat Inspection Act and the Pure Food Drug Act in
1906 – and regulations were introduced to improve the treatment of
animals before animal slaughter, to eliminate unnecessary suffering during
the killing of animals, and to make butchering, meat transport, storage
and meat processing more hygienic. These matters became subjects of
systematic research and governments' attention because better sanitation
was needed to prevent potentially very serious disease outbreaks caused by
meat-borne pathogens (such as *Clostridium, Escherichia, Listeria*) in large
urban settings.

Customer expectations had also changed. Higher incomes led first to
higher demand for meat of any kind, and the cheapest fatty cuts (belly
pork, beef shank and flank) were favored by lower-income consumers who
could finally afford substantial amount of meat. But after WW II, as the
average incomes rose and as this rise coincided with the emergence of
nutritional and health concerns in North America and Western Europe,
quality considerations became important. Demand had shifted to better
cuts, and red meat (beef in particular) began its gradual retreat as chicken
began to claim a larger share of overall meat consumption. In 1909, beef

was still dominant, with 43% share of the total meat supply, but within a few years it was surpassed by pork, and chicken demand took off during the 1960s: by the year 2000, beef accounted for 35%, chicken for 29% and pork for 27%, and a decade later chicken was on the top with 35%, followed by beef with 33% and pork with 24% (USDA 2012a).

Convenience and higher participation of women in labor force have been the two key factors behind selling more meat in portions. American chicken sales are perhaps the best illustration of this trend. In 1960, 85% of broilers were still sold as whole carcass, 13% were retailed as parts and 2% went for processing, and even a decade later the market was dominated by whole birds as the three shares stood at 70%, 26% and 4%. Then the chicken consumption took off, and by 1980, 40% of all birds were marketed in portions and 10% went into processing, directly reflecting the introduction of Chicken McNuggets by Tyson (America's leading poultry processor) and McDonald's (the top fast-food brand), and by 1990 only 18% of all broilers sold in the US were sold as whole carcasses, 56% cut up and 26% further processed (USDA 1995).

Before taking a closer look at the tempo of dietary transition and its eventual results both in Western nations and, more recently, in modernizing countries, particularly in Asia, I should describe briefly the history and impact of long-distance meat trade. This was predicated by mastering a new technique of mechanical refrigeration, a technical innovation that, by changing meat trade, changed both production and consumption in a number of countries, including such antipodal combinations as the UK and New Zealand.

Long-distance meat trade

Rising incomes brought by industrialization and urbanization created higher demand. Per capita demand for meat and growing populations intensified this trend. Britain was the first country where this combination had surpassed the capacity of domestic meat production. In 1850, average meat supply for its 28 million people was about 34 kg/capita; just three decades later, the population was 25% larger and the average meat supply could rise to 50 kg/capita because almost 40% of it came from imports (Critchell and Raymond 1912). These large-scale, long-distance shipments of live animals and chilled and frozen meat were an important new development, and subsequent technical advances have transformed it into a common and affordable form of food trade.

For millennia, the typical distance between an animal to be slaughtered for meat and people who would eat it had to be short. Absence of any forms of mobile refrigeration precluded any prolonged shipments of carcasses or meat cuts; similarly, challenges of scaling up the standard methods of meat preservation (by smoking, air drying or pickling) restricted the exports of ready-to-eat meat products to relatively small quantities. Until the introduction of first freight railway cars during the early 1830s, the only practical means of moving large numbers of live animals – most often cattle and sheep, but also pigs and horses – over long distances on land was to drive them in herds to large city markets.

The practice of herd driving (droving in Britain) is ancient: Pliny mentions flocks of geese driven from Gaul to Rome! Perhaps the two most remarkable well-documented instances of modern animal drives are that of pigs and cattle from Ohio to Philadelphia in 1815, and Leonetto Cipriani's incredible cattle drive from St. Louis to San Francisco in 1853 to supply California's gold-rush-driven demand for beef (Cipriani 1962). Cattle driving took off as a regular commercial practice only with increasing populations and higher urban incomes, and it reached its peak just before the railways reached the cattle- or sheep-rich regions during the 19th century. Relatively short drives were common around Europe, and Britain's grassy north and west and populated south led to some early long-distance droves: by 1650, Scotland was sending more than 15,000 cattle (in droves of 200–1,000 heads at just 3 km/hour) a year to the south, and cattle was also moved eastward from Wales; English droves declined even before the railways took over due to land enclosures of the late 18th century that restricted free-pass routes (Rixson 2010).

In North America, substantial long-distance cattle drives took place in Spanish Mexico and California as well as from the pre-Civil War Texas, but the peak activity was concentrated in the two post-war decades (1866–1886) when large numbers of Texas cattle were driven north to the nearest railheads to be transported to Chicago stockyards or shipped eastward (Skaggs 1973). Shawnee (Texas) trail to Sedalia in Missouri and the Chisholm trail (1,600 km from Fort Worth to Abilene and Dodge City in Kansas) were the two most traveled routes. Herd sizes varied between 1,500 and 3,000 heads of cattle, requiring 8–12 riders, an oxen-pulled chuck wagon and spare horses. Daily drives were rarely longer than 25 km, and if the animals were to arrive in reasonable shape, they needed plenty of daily rest and grazing and regular watering. Not surprisingly, losses were often substantial, but the daring, dangers and rewards associated with these drives helped to create the appeal and mystique of cowboy America (Savage 1979; Slatta 1990).

Railroad shipments of meat expanded rapidly during the second half of the 19th century (Anderson 1953; White 1993). In the US, they began in

1857 in ordinary box cars packed with ice, and the first US patent for a refrigerated car cooled by an ice–salt mixture was issued in 1867. The country's leading meatpacker, Gustavus F. Swift, began to build his fleet of well-insulated cars cooled by ice in 1878 and had nearly 100,000 units 15 years later. The first mechanically refrigerated car was patented in 1880, but ice cars remained common until WW I. In the US, the number of refrigerated railroad cars peaked in 1931 at nearly 200,000 units and then declined with the adoption of refrigerated trucks.

Live animals, mostly small numbers of breeding stock and war horses, were transported by ships for millennia. Large-scale exports of meat animals were limited due to the capacities of sailing vessels, and intercontinental transport of shiploads of animals was impractical due to often stormy (particularly in the North Atlantic and parts of the Pacific) and unpredictable voyages (including, sometimes, long spells of immobility when becalmed in subtropical high-pressure zones). Shipborne trade of live animals was thus limited to shorter distances (such as Ireland to western England, Scotland to England), and it had increased substantially only with the introduction of steam-powered vessels during the 1830s, and by 1880 the UK imported nearly 400,000 heads of cattle and almost one million sheep from North America and Europe.

Importing meat was obviously a more practical option that became possible with the advent of mechanical refrigeration (Critchell and Raymond 1912; Thevenot 1979). In 1876–1877, *Frigorifique* was the first ship that tested artificial refrigeration (using methyl-ether) on a trip from France to Argentina and back, but the first completely successful trial run on the same route came in 1878 with *Paraguay*. In 1879, *Circassia* brought the first chilled beef to London from the US, Australian shipments to the UK began with *Strathleven* in 1881 and the first New Zealand exports of meat came soon afterward in 1882 on *Dunedin*.

By 1900, there were 356 refrigerator ships (reefers) in service importing American, Argentinian, Australian and New Zealand beef, pork, mutton and lamb to Europe, and original, inefficient and costly refrigeration using compressed air was replaced by refrigerants including ammonia, sulfur dioxide and carbonic acid, with calcium chloride or ammonia circulating in cooling pipes. Britain was the largest destination of intercontinental meat trade – by 1910, imported meat accounted for nearly a quarter of the total supply of about 55 kg/capita (Critchell and Raymond 1912) – and economies of Argentina, Australia and New Zealand (they could not have made profit of their huge animal herds otherwise) were the greatest beneficiaries.

During the 1930s, chlorofluorocarbons began displacing old refrigerants (to be eventually, in the late 1980s, banned because they were destroying the stratospheric ozone), and after WW II operating costs were cut by introducing rotating compressors instead of old reciprocating (steam-powered) machines. But further expansion of dedicated fleet for shipping of refrigerated meat was checked by the introduction of refrigerated containers (reefers). They come in standard sizes (20 and 40 feet long) to fit on container ships, but their internal volume is smaller due to a refrigeration unit and ventilation, and they can keep the contents chilled, frozen or deeply frozen (to $-60°C$).

Most of the international meat trade after WW II has continued to be an intercontinental trade involving refrigerated cargoes and requiring massive cold storage in both exporting and importing countries. Global meat trade was rising steadily, and by the century's end some 15 Mt of meat (with beef and chicken each contributing about 6 Mt) were traded internationally; in 2010, the total exports had increased to nearly 25 Mt/year (or nearly 10% of the global output), with chicken accounting for nearly 40% of the total, followed by about 8 Mt of beef and 6 Mt of pork; sheep meat (mutton and lamb) added just over 1 Mt, and horse meat trade (mainly exports from Canada and South America to hippophagic parts of Europe and to Japan) has been down to less than 150,000 t/year.

Trade in live animals has also grown considerably – from less than 10 million heads in 1950 to nearly 70 million large animals (about 10 million heads of cattle, nearly 40 million pigs, and 20 million sheep, goats) by 2010 (FAO 2012) – with most of the increase due to the sales of pigs and cattle within the EU and between the US and Canada. Intercontinental trade has been dominated by sheep and goats sold to the Middle Eastern countries (Bahrain, Jordan, Kuwait, Qatar, UAE) and particularly to Saudi Arabia during the time of *hajj*. Some shipments originate in Eurasia (Russia, Romania, Turkey, China), some have come from Argentina and Uruguay but most of the animals are exported from Africa (particularly from Sudan and Somalia) and Australia.

Since 1990, Australian sheep sales to the Middle East have fluctuated mostly between 4 and 6 million heads a year, peaking at 6.8 million animals in 2001. Special ships equipped with pens are loaded with more than 50,000 animals and make the journey to the Persian Gulf in three weeks. Improved feeding, watering and care in loading have cut down the mortality rate during the transport, and in 2011 new standards specified that mortality should not exceed 2% for sheep and goats and 1% of long-haul cattle voyages; for comparison, recent performance has been as low as 0.04% for cattle on short-haul voyages of less than ten days and less than 0.5% on long trips (Australian Government 2012).

Meat in the Western dietary transition

Contrary to a common notion of a revolutionary nature of Western industrialization beginning in the latter half of the 18th century (what could be labeled a simple steam engine model of history), many aspects of the process date to the 17th century, and many of its major impacts were seen only after the middle of the 19th century (Smil 2005b). Dietary transition belongs to the latter group of changes, as even in the richest Western European countries the level and the composition of typical diets had shifted noticeably only after 1850. The US, with its low population density and abundance of wild animals, was a notable exception: pre-1850 meat consumption was considerably higher than in Europe, and, starting from this relatively high base, typical meat intakes during the closing decades of the 19th century were not that different from the levels prevailing a century later. In contrast, in Mediterranean Europe the traditional pattern of low meat intakes began to change only during the 1950s.

Dietary transition in Western Europe was preceded by the elimination of famines: after the Napoleonic wars they belonged, with the obvious exception of Ireland, to history. The best available statistics show that between 1860 and 1900, average daily food supply in all major countries (France, Germany, UK, Italy) rose to 2,700–3,300 kcal/capita, far above any risk of famine, while in the poorest regions and among the poorest populations, malnutrition continued well into the 20th century (Grigg 1995). A relative wealth of historical data allows a fairly reliable reconstruction of rising meat supply in France and Britain during the 19th and the early 20th centuries (Dupin et al. 1984; Perren 1985).

French meat supply (carcass weight) stagnated during the first half of the 19th century at less than 25 kg/capita, rose to about 40 kg by the year 1900, fluctuated between 1914 and 1945 from 45 to 55 kg and reached about 55 kg by 1950; by 1975, it was close to 100 kg, and it ended the century roughly at the same level. Major shifts between 1950 and 2000 were in accord with composition changes elsewhere in the West, including first an increase but then a stagnation and decline of red meat consumption (a result of *la guerre aux matières grasses*) that was more than compensated by steadily higher purchases of poultry (tripling in four decades before 2000) and processed meats, whose per capita intake nearly quadrupled between 1960 and 2000 (Monceau et al. 2002).

British per capita supply (also as carcass weight) rose faster than the French rate during the latter half of the 19th century, roughly tripling the 1800 level to a fairly high rate of nearly 60 kg by the year 1900 (Perren 1985). During WW I, it fell by more than 25% compared to the pre-war years, recovered and stagnated during the 1920s and 1930s (averaging just over 61 kg, including poultry and game, between 1934 and 1938) before falling to nearly 45 kg/capita during and right after WW II and then recovering to more than 60 kg by 1950. Interestingly, the post-war rationing (lifted only in 1954) provided only 20% less meat than the British consumer bought freely a generation later (Hollingsworth 1983). Subsequent growth pushed the mean above 70 kg/year by 1970, but the total did not rise above 80 kg/year by the year 2000.

German meat supply rose from less than 15 kg right after the Napoleonic wars to almost 25 kg by 1860, by 1892 it was 32.5 kg and during the first five years of the 20th century it was almost 47 kg/capita (Abel 1962). Wartime declines were even more pronounced than in the UK, but the post-WW II rise of meat consumption was rather rapid, bringing the per capita mean (in carcass terms) to more than 60 kg/year in 1960 and to the peak of about 96 kg in 1990; the mean then declined to 83 kg/year in 2000 when it was nearly 15% lower than in France, 25% lower than in Spain and a third lower than in Denmark. In Russia, average annual meat supply in 1913, the last peaceful year before the upheavals of WW I, the Bolshevik Revolution and a civil war, was 29 kg/capita, and it took more than 50 years to double that rate (TsSU SSSR 1977).

Although the traditional Mediterranean diets were not vegetarian, they contained only a small amount of meat, with more animal protein coming from fish and dairy products. Typical pre-WW II annual rates were lower than 15 kg/capita, and a detailed study of traditional diet on Crete showed that in 1949 the islanders derived only 4% of their food energy from meat and that their total annual intake of meat, fish and eggs was only 23 kg/capita (Allbaugh 1953). Until the late 1950s, the mean meat supply remained below 20 kg not only in Greece and Portugal but also in Spain. Following increases, accelerated by new economic prosperity brought by accessions to the EU (Greece in 1981, Spain and Portugal in 1986) raised the mean supply rates (in carcass weight) in 2000 to about 90 kg/capita in Greece and Portugal and to 112 kg/capita in Spain. The Spanish meat supply had more than quintupled between 1960 and 2000, and in the EU it is now surpassed only by Denmark, Luxembourg and Cyprus, the last divided country being yet another example of a rapid demise of the traditional Mediterranean low-meat diet.

The US history of meat consumption differs from the European experience because the abundance of game and little land constraint on production of feed allowed generally higher per capita supplies, although dietary differences due to social status remained obvious. For example, during the mid-18th century, well-off New Englanders consumed close to 70 kg of meat a year (Derven 1984), as much, or not much less, than many of their early-21st-century descendants. Most of it was fresh meat, while poorer Americans (as well as slaves) ate smaller amounts of mostly salted meat and hunted more often.

Average US meat consumption was well ahead of even the highest European means throughout the 19th century, and only after WW II did the supply in some European countries rise close to the US level. Economic prosperity that followed the Civil War, expansion of Western grazing and Midwestern corn production, introduction of refrigerated railroad meat transport in 1870 and the introduction of the pressure cooker in 1874 (allowing easier preparation of tougher cuts of meat) combined to produce a substantial increase of meat consumption, mostly beef and pork. By the late 1880s, Atwater (1888, 259), an early student of American diet, concluded that Americans were "eating in excess of energy demands, with excesses of meats, especially of the fatty kinds." By 1890, the American diet (also full of sugar and starch) was not unlike the pattern that prevailed a century later.

A well-documented record of America's food supply begins in 1909 (USDA 2012a). During that year, the mean per capita availability was equivalent to about 51 kg of boneless trimmed (edible) weight, with red meats supplying about 90%. This would be roughly 75 kg of carcass weight, and adding the excluded internal organs and wild meat would raise the total to about 80 kg/capita. Annual supply equivalent to about 50 kg/capita of boneless meat changed little during the next two decades, but during the Great Depression it declined to 40 kg by 1935. Wartime economy brought rising incomes and higher meat consumption (62 kg of red meat in 1942), and meat rationing (between March 1943 and November 1945) was hardly a hardship at 59 kg of red meat a year per capita (Bentley 1998). In contrast, GI meat rations in the US Army were an equivalent of 106 kg of red meat a year, and in the Navy were much higher, at 165 kg possibly the highest meat rations in history.

With meat supply already relatively high, the post-war prosperity did not make that much difference: in 1950, the mean supply of red meat and poultry was just 6% above the 1940 level, and by 1960 the difference was still less than 20%. Boneless red meat consumption continued to rise during the 1960s, and it peaked at 61.5 kg in 1971; subsequent slow decline brought it to less than 52 kg by the year 2000 and to less than 48 kg by 2010 (USDA 2012a). Availability of poultry had doubled between 1950 and 1970 (to 22 kg/capita), and by the year 2000 it had surpassed 30 kg/year, growing further to nearly 34 kg by 2005 before declining slightly by 2010. Total supply of boneless meat reached about 76 kg/capita by 1980, remained unchanged a decade later and by the year 2000 it rose to just short of 82 kg/capita and then declined slightly to less than 80 kg during the first decade of the 21st century. In terms of carcass weight, the US meat supply rose from about 75 kg in 1950 to roughly 120 kg/capita by the year 2000 (USDA 2012a).

Transitions in modernizing economies

Since the 1950s, most countries – with notable exceptions of those engaged in protracted civil wars (Sudan being the worst example) or ruled by regimes willing to starve their own people (such as North Korea with its recurrent famines) – have seen significant economic advances that were accompanied by gains in average quality of life and often by impressive improvements of food supply. Another notable exception has been India, a nation with major culturally dictated restrictions on meat eating that has also struggled with providing enough food for its expanding population: it has done so only by maintaining an overwhelmingly vegetarian diet, with average annual meat supply remaining at only little more than 3 kg/capita.

But in many other populous countries, the combination of very low traditional meat intakes and of nearly universal human propensity for consuming more meat as incomes rise has translated into average annual per capita meat availabilities that are often multiples of pre-1950 supplies. In Brazil, the increase has been roughly fourfold, to a very high rate of about 80 kg/capita thanks to two quintessential dishes of Brazilian cuisine, *churrasco* (grilled meat, mostly beef) and *feijoada* (stew with beans and meat, originally the cheapest cuts of pork, now with pork, beef or ham). Indonesia has also seen a quadrupling of meat supply but only to little more than 10 kg/capita; Mexico's meat availability has tripled to more than 60 kg/capita, and South Korea has expanded its average meat supply nearly 20-fold, to almost 50 kg/capita. Only Africa's populous nations have seen relatively small gains: a doubling in Egypt to over 20 kg/capita and a 50% gain in Nigeria to still less than 10 kg/capita.

But the two most important instances of a rapid dietary transition accompanied by much increased meat consumption have been those of China, the world's most populous nation and now also the world's second largest economy (in aggregate terms), and Japan, now the world's third largest economy and Asia's most technically advanced nation. Traditional cultures of these two nations have a great deal in common (the younger one being a major derivative of the older) and have shared many foodways – but their recent history has been very different and some food preferences have been highly idiosyncratic: Japan with its resolute choice of modernity starting in the late 1860s, subsequent rapid economic rise, more than millennium-long ban on eating meat and the world's highest fish consumption; China with its first three post-imperial generations (1911–1976) lost to instability, civil war and Mao's brutal rule, with its belated but intense post-1980 modernization drive and with its traditional preference for pork.

China's modernization began slowly in 1980, four years after Mao Zedong's death in 1976, accelerated during the 1990s and it has produced one of the highest periods of economic growth rates and one of the fastest dietary transitions in history. As already noted, meat consumption by most of China's population (poor subsistence peasants) remained very low during the first decades of the 20th century and that traditional pattern had changed little during the first 30 years of the Communist state established in 1949, and during the greatest man-made famine in history (1959–1961), average food supply deteriorated so much and so widely that more than 30 million people died and hundreds of millions were malnourished (Becker 1998; Smil 1999).

In 1961, average per capita food energy supply was below 1,500 kcal/day and annual meat supply was less than 4 kg/capita. By 1970, daily food energy availability was still below 1,900 kcal and meat supply remained below 10 kg/year. In 1978, two years after Mao's death when Deng Xiaoping came to power, typical peasant diet differed little from the pre-famine pattern in 1957: preferred fine grains (highly milled rice and white flour) accounted for only half of staple carbohydrates, and pork-dominated nationwide meat output was only about 7.5 Mt, prorating to less than 12 kg/capita. The subsequent dietary transformation was so rapid that just four years later in 1982, even peasant families consumed three-quarters of their staple grain as milled rice or wheat flour, a ratio enjoyed previously only by privileged urban residents on special rations.

In 1980, the slaughter of 210 million pigs with an average live weight of 89 kg resulted, for the first time, in a nationwide pork supply of more than 10 Mt. By the spring of 1981, the Chinese media noted the end of pork supply shortages, and the rising output brought the average per capita availability to more than 12 kg/year and to nearly 20 kg/year in large cities. By 1990, per capita meat supply rose to about 26 kg and then it nearly doubled during the 1990s to almost 50 kg/year, with pork accounting for nearly two-thirds and poultry for more than one-fifth of the total. Subsequent increase was small as per capita meat supply remained below 55 kg/year, higher than in Japan.

Two generations after its re-opening to the world, Japan's annual meat supply remained very low: in 1900 it prorated (in terms of carcass weight) to just 800 g/capita (most of it beef), by 1925 it was about 1.7 kg (half beef, 40% pork, 10% horse meat) and in 1939 it was almost exactly 2 kg/capita, that is, still only about 5 g/day. In practice, it meant that most people ate small portions of meat only a few times a year. The war lowered even those tiny rations, and by 1950 the carcass weight of domestically produced meat was still below the low level of the late 1920s. Rapid increases of domestic meat output began only during the mid-1960s, and it was accompanied by rising meat imports: this had nearly tripled the annual per capita availability between 1965 and 1985, from 12 kg to 35 kg, and by the year 2000 the mean supply was nearly 45 kg/capita, or about 29 kg in actually edible portions, nearly a sixfold rise since 1960 – but even so the Japanese still prefer to eat more seafood than meat (Matsumura 2001; Smil and Kobayashi 2012).

Unlike populous China and land-short Japan, Brazil has always had ample grazing land (natural or created through deforestation) and hence a high meat-producing potential. Unfortunately, since the 1970s it has expanded its traditionally substantial beef production through extensive conversion of Amazonian forests to pastures, a trend that has only recently seen some encouraging slowdown (Nepstad et al. 2009). Among the world's most populous nations, only India, an ancient vegetarian society, has shown a minimal shift toward even moderate meat eating (Hopper 1999). Pre-WW II food balances indicate average annual meat supply of about 6.5 kg/capita, a rate that was not surpassed until the mid-1960s; but FAO's food balance sheets show average supplies below 4 kg both in the early 1960s and in the year 2000.

During the early stages of economic modernization, there is a fairly high correlation between average per capita meat consumption and wealth (be it measured by GDP or disposable income), but there is an obvious reduction in the increase rate once the GDP surpasses $10,000. And there are also many notable outliers (explained largely by cultural and environmental differences), with some countries reaching per capita consumption plateaus at much lower rates than others as is clear by comparing Turkey and Brazil or Japan and Spain. And although the post-WW II decades of growth have created expectations of continuing increases or, at least, leveling off at relatively high consumption levels, it is highly likely that the combination of population aging, health concerns and economic hardships will bring declining average per capita meat intakes in many countries.

Globalization of tastes

More meaty diets have become a universal component of dietary transition, and while many old cultural biases and nutritional preferences have changed or have entirely disappeared, others are too ingrained to be changed in a matter of few generations. In terms of the global meat consumption, the most consequential of these traditions are the food taboos and dietary preferences of Islam and Hinduism. Muslim proscription of pork prevents a billion or so people from consuming the most affordable of all red meats, and the Hindu avoidance of bovine meat (except for people traditionally associated with the slaughter of cattle and water buffaloes and with the processing of leather) makes a lasting imprint on the dietary habits of India. And with the exception of Japan, Latin America and half a dozen of hippophagic nations of Europe, there is a continuing widespread abstention from eating horse meat.

And while smaller affluent nations that cannot satisfy meat demand of their populations through domestic production can afford to make up the difference by imports (best examples include Singapore, South Korea and the Persian Gulf states), that option has been largely foreclosed for nearly all poor populous nations. Those countries face two kinds of limits: environmental limits precluding further expansion of domestic meat production and shortages of funds to import meat or live animals. The environmental limits are particularly true as far as beef is concerned as the countries with limited area of suitable grazing land (or with already overgrazed grasslands) and with the necessity to use scarce farmland for food production (rather than for growing concentrate feed) cannot resort to either of the two beef-producing strategies and will prefer to spend their limited foreign exchange on much less expensive frozen poultry rather than on chilled beef (leading examples in this large category are most countries in sub-Saharan Africa).

These caveats are necessary in order to understand the true meaning of this section's title: it should not suggest that widespread meat-eating preferences have become completely global (as many poor populous countries have been obviously prevented from joining the shift toward eating significantly more meat) but rather that their remarkably rapid diffusion and adoption has now spread to all continents and either to a majority of the world's countries or to a large share of populations. Undoubtedly, the most notable expression of globalized tastes, indeed of the process of economic globalization in general, has been the fact that the world's two largest multinational fast-food chains have based their success not on starchy staples or artfully prepared vegetables but on affordable servings of meat.

The post-WW II expansion of poultry production led to the rise of one of the world's most successful and truly global enterprises, Kentucky Fried Chicken (KFC), now a part of Yum! Brands (KFC 2012). By 2010, the company had some 5,200 branches in the US and more than 15,000 restaurants in 109 countries on all continents, with a large-scale presence in Asia (operating in Japan, South Korea, China, Indonesia, Philippines, Thailand, India, Pakistan, Turkey and all countries of the Arabian Peninsula). Besides the poorest parts of Africa, its only notable absences are Italy and Scandinavia in Europe, and Bolivia and Argentina in the Americas.

KFC has plenty of local competition in all major markets – from Australia's Chicken Treat to Nigeria's Tastee Fried Chicken and from Saudi Al Baik to English Chicken Cottage – but its greatest global rival is the company that sold only hamburgers for the first 30 years of its existence: chicken meals now account for an increasing share of its offerings. The company was founded in 1948 and it introduced Chicken McNuggets only in 1979 and McChicken in 1980, but within a few years the company's worldwide chicken sales were second to only KFC. In 1995, when McDonald's opened its operation in the world's most populous cattle-venerating country, it had to rely primarily on chicken supplemented by fish, *paneer* and veggie burgers. Indian offerings now include standard McChicken and Chicken Maharaja (with double filling of breast meat) and McSpicy Chicken (leg meat) and BigSpicy Chicken Wrap to cater to Indian tastes.

But McDonald's has become the world's largest purveyor of fast meat – and perhaps the most recognizable symbol of the globalization of taste – thanks to its offerings of fried beef served in assorted sizes of plain and variously flavored hamburgers (McDonald's 2012). The company began its foreign expansion in 1967 with franchises in Canada and Puerto Rico, and in 1971 it opened its first franchise in Japan and now the country's nearly 4,000 franchises of *Makudonarudo hanbāgā* (the world's second highest count compared to about 14,000 in the US and about 1,100 each in Canada, UK and Germany) compete not only with Burger King and Wendy's but domestic hamburger chains led by MOS Burger, Freshness Burger, First Kitchen, Lotteria and Sasebo Burger (McDonald's Japan 2012). In 1967, the company introduced Big Mac; the Quarter Pounder followed in 1973.

In the US, the company fought and won "hamburger wars" of the 1970s with Burger King and Wendy's and in 1984 opened its first restaurant in downtown Beijing. Burger King came to China late, in 2005, and both companies have to work hard to overcome the Chinese beef bias and preference for chicken. By the end of 1996, Big Macs were on sale in 100 countries; by 2011, the number approached 120 in nearly 33,000 franchises. But the overall US beef consumption is so large that the company still buys only about 3% of the country's beef supply. Burger King is the world's second-largest burger chain with more than 12,000 franchises in more than 70 countries.

Even as the US beef consumption has been declining, the burger companies have been introducing more meaty and more fatty choices: McDonald's regular hamburger contains 170 g of meat, 570 kcal and 30 g of fat (a heavier Big Mac, 214 g, is leaner and has 540 kcal), while Burger King's Double Whopper with cheese weighs 398 g and has 990 kcal and 64 g of fat and a Tripple Whopper with cheese at 480 g delivers 1,230 kcal and 82 g of fat (Acaloriecounter 2012). Many record-breaking attempts have come up with obscene creations containing first more than 50, then 100 and eventually more than 300 kg of meat. Another global fast food, pizza, is also strongly tied to processed meat. Almost anything has been put on pizza (from anchovies to whitebait) – but in America, pepperoni has been the favorite topping, with more than 100,000 t/year put on nearly 40% of all pizzas (Yourguidetopizza 2012). More than 60% of all pizzas served in the US contain some meat product, with ground beef being a favorite pizza topping in Saudi Arabia as well as in the Netherlands, and minced mutton in India.

Output and Consumption: Modern Meat Chain

Even when modern animals are under the best possible care and when they benefit from optimal feeding and disease prevention, their rate of meat production remains circumscribed by the specific requirements of their reproduction, growth and maturation. All of these variables have been manipulated and altered by modern livestock husbandry: reproduction through artificial insemination of mammals and controlled incubation of birds and the post-weaning growth in mammals and post-incubation development in poultry have been accelerated by combining confinement (reducing energy needed for physical activity), disease prevention (eliminating slower growth due to infectious and digestive disorders) and balanced feeding (delivering optimum balance of nutrients). But there are practical limits to contracting the span between birth and slaughter, and these metabolic imperatives will always determine the differences among the production rates and costs (financial, energetic, environmental) of different meats.

Animal slaughter in traditional societies (in pastoral communities as well as in pre-industrial villages) was an infrequent (and often festive, memorable and greatly anticipated) affair with the killing done by throat cutting. Hundreds of millions of ruminants, pigs and poultry are still killed annually by pastoralists and subsistence farmers who use nothing more than a sharp knife and, if needed, some help in holding or hobbling an animal. A high degree of carnivory in modern societies has transformed animal slaughter into an impersonal mass-scale industry that operates at truly astonishing rates – but whose methods of killing, when properly deployed, are more humane than are the ways of traditional slaughter.

Large carcasses hanging from hooks are not a sight repeatedly seen by modern Western meat eaters as butchering of animals has largely retreated from the point of sale to the place of slaughter or to associated specialized meat processing plants that prepare a wide variety of conveniently packaged cuts. Modern meat preparation also differs from traditional ways by the degree to which it uses raw meats to prepare a large range of processed meat products, a category best described by the French term *charcuterie* that embraces hams and salami. Moreover, a high degree of processing also applies to non-edible meats and fats that are turned by specialized facilities into animal feed, pet food, fertilizer and other industrial products.

The single most important innovation that has transformed modern meat retail has been the evolution of an unbroken chain of refrigeration as chilled carcasses from slaughterhouses move in vans, wagons or ships to processing facilities or large-scale retail stores and as packaged cuts processed meats move in chilled distribution trucks and vans to retail shops to be taken to household refrigerators and freezers. The only time modern meat is not chilled or frozen after slaughter is when it is carried or transported from a shop to a home. Naturally, this ubiquitous refrigeration has drastically lowered meat spoilage and reduced overall distribution, retail and household meat loss – but meat waste at household level, a part of unacceptably high post-production food losses throughout the affluent world, remains excessive.

Accounting for these meat flows is done in all of the just noted categories along the production chain. There are statistics showing the average take-off rates (percentages of respective herds or flocks that are slaughtered in a given year) and average live weights and carcass weights; those summing up the heads of annually slaughtered animals and the equivalent masses in carcass terms and in retail cuts, with the latter category sometimes separated into take-home meat with bones in and trimmed, boneless cuts; and, infrequently, those quantifying actual food intakes at the household level. Naturally, the total quantities are diminishing as the accounts progress along the livestock-table chain. Unfortunately, comparisons (historical or international) are often made without specifying the category they use or, worse still, they misleadingly contrast two different categories (commonly per capita consumption in terms of carcass weight for one country with per capita supply in terms of trimmed, boneless cuts for another). I will sort out these discrepancies in a clear sequential manner.

Changing life cycles

The life cycle of domesticated animals imposes the most fundamental limit on meat production. With mammals there must be, once they reach sexual maturity, the sequence of estrus, breeding, gestation, lactation and nursing, followed by weaning, recovery of females to begin a new breeding cycle and growing period during which the young animals reach desirable slaughter weight. Once birds reach sexual maturity, females lay a clutch of eggs, incubate them and hatchlings must grow to adult size. Some of these life cycle segments have been considerably affected by selection, feeding and overall management of domesticated species, and this shortening of life cycles has been a key to higher productivity of meat animals.

These changes have been most obvious for chicken, and enormous expansion of the US broiler industry after WW II would have been impossible without shortening or at least streamlining and optimizing (as with artificial incubation) every stage of the bird's life cycle. These changes brought about the systematic innovation and institutional transformation that created the very paragon of modern agribusiness: integrators purchase breeding stocks, hatch chicks and transfer them to growers, then harvest them for slaughter, with all operations optimized and their duration reduced to unprecedented minima (Sawyer 1971; Boyd and Watts 1997).

Chicken's wild ancestor, the red jungle fowl (*Gallus gallus*) of South Asia, reaches sexual maturity 25 weeks after hatching but for domesticated varieties that span is normally 20% shorter (20 weeks), with minima down to 18 weeks, while their normal clutches are roughly twice as large, averaging a dozen eggs. Wild fowl (omnivorous foragers) would take up to 6 months to reach maximum weight (their normal life expectancy is at least 6 years), modern free-range chicken reaches slaughter weight in about 14 weeks and chicken fed grain-based rations in confinement are market-ready in less than 6 weeks although they weigh 50–60% more than their free-ranging predecessors.

Reliable US data allow us to trace these weight gains since the mid-1930s when the average slaughter weight was just 1.3 kg; in 1950, young chicken averaged less than 1.4 kg, by 1975 the weight was up to nearly 1.7 kg, in 2000 it surpassed 2.25 kg and in 2010 it was 2.58 kg, almost a 90% increase in 60 years (USDA 2011). And it has taken less and less time to reach those rising market weights: 113 days in 1935, 95 days in 1945, 47 days by 1995 and no more than 42 days by 2010. As the birds feed essentially *ad libitum*, this substantial shortening of feeding period has more than doubled average feed conversion ratio, and this remarkable gain in feeding efficiency will be addressed in detail (and for all meat animals) in Chapter 4.

The net result is that by the late 1990s, a broiler reached nearly twice the weight than the bird from the 1920s; even more impressively, by 2001 the modern broiler was nearly five times as large at 42 days of age as its predecessor in 1957, and 85–90% of this change in growth rate was due to genetic selection and the rest to better nutrition (Havenstein 2006). No wonder that production of chicken meat has been growing faster than that of any other kind of meat, and this growth is far from reaching a plateau. Worldwide output of chicken meat rose from just a few million tons in 1950 to about 16 Mt by 1975 and to 59 Mt by the year 2000, and a decade later it increased by about 50%.

These achievements may be admirable when judged in dispassionate productive efficiency terms, but they rest on a combination of aggressive interventions, all evolved to support the most important condition, that of intensive confinement. Growing birds could be confined in limited spaces only after it was discovered that the addition of vitamin D (in cod liver oil) prevented leg weakness that resulted from the absence of exposure to outdoor UV light (Hart et al. 1920). Artificial incubation in hatcheries operating under optimized environmental conditions (made possible by the post-WW I electrification) was the other prerequisite as it assured continuous supply of new birds for brooding.

Similarly, selection, confinement and optimized feeding accelerated the production of pork. A female hog is ready to reproduce about 32 weeks after her birth, and gestation lasts about 16 weeks (114 days) after mating or artificial insemination. Sows farrow (give birth) to litters of 7–12 (average 9) piglets, each weighing about 1.5 kg. Piglets are nursed for two to three weeks, and sows can be ready to reproduce shortly after weaning and hence they average slightly more than two litters a year. Those are natural constants, but breeding for meat growth and feeding in confinement have greatly shortened the post-weaning growth. In wild, the normally active pigs – foraging (rooting) in herds and covering considerable distances every day – reach full adult weight only after 2 years for females and 3 years for males and can live for as long as 8–15 years. Pigs grown in semi-confined conditions and fed *ad libitum* obviously mature faster (in less than a year) and those reared in severely confined conditions and fed optimal diets can reach their slaughter weight of 110–120 kg in as little as 22 weeks after their birth but 24 weeks, or about 5.5 months, is a more common span.

Cattle cycle is naturally the longest of the three dominant species of meat animals, and its greatest change came only during the post-weaning stage. Heifers (young females before their first pregnancy) reach sexual maturity 15 months after birth but are mated or artificially inseminated to have their first calf at the age of 2 years. Their gestation period lasts nine months, twin calves are a rarity, calves stay with cows (even in the US the pairs are maintained on pasture with minimal or no grain feeding) six to eight months and their diet may be supplemented by grass. After weaning, most male calves are castrated (these steers are then fed to reach market weight), and only a small share of all males is kept to become breeding bulls. Cows are re-bred 2–3 months after their birth resulting in a 12-month calving interval, and they stay productive in a breeding herd for at least 7 years.

Heifers and steers could be left on pasture, and it could then take at least two but as much as four, even five, years before they reach their slaughter weight. But the most common US practice is to put the weaned animals on grass or to feed them other roughage for 6–10 months until these stockers reach at least 270 kg (and as much as 360 kg), and they are then moved to feedlots where they are fed grain-based rations supplemented by a minimum of roughage required by ruminants. After five months of intensive feeding, they reach their market weight in excess of 500 kg. Consequently, the standard US sequence from the birth of a female to the slaughter of its offspring (heifer or steer) that has gone through stocker and feeder stages is at least 50 and up to 56 weeks compared to at least 145 weeks for animals raised solely by grazing.

In practice, these differences are easily traced by looking at average take-off rates, that is, the ratio of animals slaughtered in a year and their total number in a herd or in an entire country. Take-off rate of one means that it takes, on average, one year for that species to reach slaughter weight. Obviously, take-off rate calculated by using the entire stock of a particular species and including all breeding animals as well as those that never reach the slaughter age will be lower than those using just the numbers of animals that were reared for meat and survived into maturity: that is why, for example, the take-off for US pigs is now about 1.42 or 36 weeks rather than just 24 weeks when including only the animals finished by intensive feeding.

Traditional pig breeding was well below that rate, with take-off rates as low as 0.4 (i.e., pigs slaughtered only when more than two years old, a rate typical for China until the 1960s); modern intensive pig production has take-off rates in excess of 1.5, while the average take-off ratios for cattle are now about 0.38 for the US and 0.32 for the EU, implying average slaughter ages of 2.6 and 3.1 years (again, these spans are considerably longer because they include all sire and dam animals and all animals that are culled before reaching maturity). Average take-off rates in modern intensive chicken production are as high as 5 (or 10 weeks including the total bird stock), while in China (with a mixture of modern and traditional feeding methods) the mean is still a bit below 3, or an average of 18 weeks from hatching to slaughter.

Slaughtering of animals

In 2010, the global slaughter of large mammals killed for meat surpassed 300 million heads of cattle and water buffaloes; it approached 1.4 billion pigs and 1 billion for sheep and goats, and nearly 5 million horses and more than 2 million camels were also killed. Numbers for poultry are less accurate, but the best totals are about 55 billion chicken and more than 3 billion ducks and turkeys (FAO 2012). The US totals in 2010 were 34 million heads of cattle, 109 million pigs, nearly 2 million sheep and goats, 8.6 billion chicken and more than a quarter million turkeys and ducks (USDA 2012b). For comparison, the totals for 1900 were 10.8 million heads of cattle, nearly 52 million pigs and 12 million lambs and sheep. Daily and per capita rates are perhaps even more impressive: in 2010 in the US nearly 100,000 heads of cattle, 300,000 pigs and 24 million broilers were killed every day, and nearly 30 animals (more than 98% birds) are killed per capita every year.

Buddhists see this as taking lives of sentient beings on an unprecedented scale, Animal Liberation Front calls for a violent response, earnest vegans are abhorred by the killings and urge more soy-based diets – while agricultural economists extol the aggregate value added, lobbyists do their utmost to keep the overall levels of meat supply up, and billions of consumers enjoy satisfying protein- and lipid-rich meals. All of these sentiments aside, killing on such a massive scale is a major logistic challenge: an occasional slaughtering of a solitary animal in a shed or in the open is one thing, killing hundreds to thousands of animals a day under one roof is a very different proposition. Moreover, killing of such a large animal as a grown steer – in 2010, its average slaughter weight in the US was nearly 600 kg – is not an easy task, and neither do smaller animals (pigs, sheep and goats) go to their deaths as lemmings.

Before getting killed and bled and butchered, they must be segregated, gathered, loaded for transport to slaughterhouses, led to killing sites and restrained. If we believe that animals can suffer, then the least that could be done is to minimize, if not eliminate, many potential causes of such suffering that are encountered throughout an often lengthy process that precedes their deaths (Blackmore 1993). That is why modern societies have in place legal norms designed to assure humane methods of slaughter. The term humane slaughter is, of course, a stark *contradictio in adiecto* (would not a truly humane treatment dispense with killing?) but, at the same time, it is the minimum that should be done by any carnivorous society.

Killing without stunning remains the norm in ritual Muslim (*zabiha*) and Jewish (*shechita*) slaughter required to produce *halal* or *kosher* meat, but in modern mass slaughtering of meat animals, stunning, a procedure to cause immediate loss of consciousness, always precedes killing (FAO 2001; Shimshony and Chaudry 2005; Grandin 2006). In the US, *The Humane Slaughter Act* of 1958 was first applicable only to livestock slaughtered in plants supplying the US federal government, and it was only in 1978 when *The Humane Methods of Slaughter Act* made the requirements applicable to all domestic facilities as well as to foreign slaughterhouses exporting meat to the US (Thaler 1999). Poultry has been excluded from these acts because electrical or CO_2 stunning was already in widespread use during the 1950s. In the EU, slaughtering is regulated by the Council Directive 93/119/C (CEU 1993).

Slaughterhouses now always use restraining devices – to which animals should be driven in a single file, quietly, without any visual distraction and without forced prodding, and that should have non-slip floors and appropriately engineered restraints devoid of sharp edges – before stunning the animals. Grandin used the understanding of livestock behavior to design optimal facilities for animal handling that would guarantee humane treatment of cattle and pigs in slaughterhouses supplying McDonald's (Grandin and Deesing 2008).

Stunning must induce unconsciousness that lasts long enough to prevent any potential recovery before an animal is killed, and it is done in three different ways: mechanically, by a captive bolt shot from a pistol placed at appropriate sites on the forehead of cattle, horses and pigs, and top of the head of sheep and goats; by electric current (requiring low-voltage AC with minima of 1.5 A for cattle, 1.25 A for pigs and 120 mA for broilers and layers and inducing epileptic state in the brain); or (since the early 1950s) by exposure to gases (very high levels of CO_2 or inert gas mixtures, Ar or N_2, for pigs, and mixtures of CO_2, N_2 and O_2 for poultry). Effective ways of electric stunning were developed for cattle in New Zealand, including the Wairoa process (head-only stunning with heart still beating) that is generally accepted as *halal*, and electrical stunning is the most widely used method in the EU, with some slaughterhouses preferring head-to-chest stunning rather than the head-only arrangement.

Electric stunning is commonly applied to pigs, sheep and poultry. For mammals, the current is applied by two electrodes using tongs firmly emplaced on either side of their heads or under the jaw and on the neck behind their ears: because this head-only stunning is reversible, animals must be bled immediately before regaining consciousness. Electric stunning of birds is done most commonly by hanging them by their feet and passing their heads through electrified water bath. Boyd (1994) argued against the use of electrical stunning devices for poultry, mainly because he found that in automated water-bath stunners a proper stun was not achieved "with alarming frequency," and Raj (2006) noted that the amount of current applied to birds varies widely and that rational stand-ardization of the process is needed.

Mechanical and electric stunning induce unconsciousness within less than 2 ms, that is, in time far shorter than is considered, from the sensory perspective, as instant (less than 300 ms). In contrast, gas stunning/killing (using inert gases or CO_2) is not unsuitable for large animals, and the onset of anoxia is obviously accompanied by respiratory distress. Killing after stunning is done by severing jugular veins and carotid arteries and ensuing bleeding (exsanguination) that causes cerebral anoxia: constantly sharpened knives and practiced skills are all that is needed at that point. Unfortunately, things may not go perfectly at any stage of this process (Grandin 2000, 2002) or, as Blackmore (1993) noted, euthanasia (good death) is not always *eu*.

Social animals, fearing separation and forced removal, may be difficult to load onto crowded transport trucks, wagons or ships; they may balk when led to slaughter and often may experience more stress and pain *en route* (being prodded, dragged, forced to walk over the top of other animals, thrown or poked in sensitive areas) than during the actual killing. Animals transported in containers should be killed as soon as possible after their delivery and not left without water or feed for many hours. Before they are stunned, live animals may be exposed to stunning or killing of their kin. Stunning procedures may work imperfectly due to inaccurate targeting and insufficient velocity and size of captive bolts. Animals can receive accidental pre-stun electric shocks, and inadequate current and voltage can lead to recovery of consciousness. And gas stunning does not produce an instant insensibility as aversiveness to high CO_2 concentration can cause severe respiratory distress.

Grandin's (2006) key conclusion is that properly designed restraining and stunning equipment must be operated by well-trained employees, and that the use of objective numerical systems that McDonald's began to use in 1999 in auditing US beef and pork slaughter plants (based on animal vocalization during handling and stunning) can lead to substantial improvements. The best possible performance is also in the interest of slaughtering facilities because stresses experienced by animals may show in poor quality of meat due to loss of muscle glycogen and resulting in elevation of pH (5.9 compared to normal 5.6) that darkens the meat's color (dark cutting beef), affects taste and hastens spoilage, and in lean meat that is blood-splattered from burst vessels and is totally unsuitable for retail (Hanson et al. 2001).

Dead cattle first has its head removed, hide stripped, shanks removed at knees and hocks, the tail cut off, the carcass split and all internal organs taken out (some post-bleeding operations may take place in a different order). Killed pigs are bled, scalded and de-haired by scarping; head, leaf fat and viscera are removed but shanks and skin are left on. Broilers are gathered (usually by hand) from the growing-out houses, put into boxes and driven to nearby processing plants. There they are hung by feet on conveyor belts and stunned in vats of electrified salt water, and neck cutters then sever their carotid arteries. Birds are then bled, scalded and mechanically de-feathered, heads and feet are cut off, carcasses are manually eviscerated, inspected, washed, chilled, packed and labeled, and all the time their internal temperature should be less than 4.4°C in order to minimize the risk of *Salmonella* infestation (USDA 1999).

Muscles of large animals become meat only after their glycogen is decomposed to lactic acid and the muscle pH falls from neutral at slaughter to 5.8–5.4 a day later. This lower pH imparts typical meat taste and flavor, and it also helps to extend storage life by retarding growth of bacteria. In stressed animals, particularly in pigs, pH can drop too rapidly to 5.8–5.6 (and eventually too far to 5.0) while the carcass is still warm, resulting in what is known as PSE (pale, soft, exudative) pork oozing moisture and resulting in excessively dry cooked meat. The reverse of this condition, most common in beef, is dark, firm, dry meat whose pH stays above 6.0 because a fatigued animal exhausted its stores of glycogen; it has a sticky feel and is tough and tasteless when cooked.

The ratio of carcass weight and live weight for all animals of the same species slaughtered in one year gives the average dressing percentage. Naturally, these shares vary not only with species but also with different breeds and styles of processing and are affected by gut fill (after a day of fasting, the dressing share will be up to 5% higher than right after full feed) and muscling and fatness (well-muscled as well as fatter animals will have a higher dressing percentage). Dressing percentage is the weight of chilled carcass as a share of live weight. In the US, carcass yields are just above 60% for beef cattle (and 59% for dairy steers), about 75% for hogs dressed packer-style (head off, leaf fat and kidneys removed, skin and shanks on) but only 60% for piglets, and a bit above 50% for market lambs (the ratio for goats is similarly low); in 2010, the actual exact nationwide averages for US beef cattle, hogs and lambs were, respectively, 60.3%, 74.7% and 51.1% (USDA 2011). Typical dressing shares range between 62% and 72% for chicken (with 65% being a good ready-to-cook average for young birds) and 78% for turkeys.

All American meat animals have been getting bigger: average live weight of beef cattle was 434 kg in 1950 compared to just over 580 kg in 2010 (34% gain in 60 years), and during the same period average dressed weight went up by more than half, from 225 kg to 350 kg; for hogs the live weight and dressed weight gains were, respectively, 14% and 48%, rising to 124 kg and 93 kg in 2010 (USDA 2011). Much as obese Americans have been putting new stresses on transportation infrastructure, heavier cattle has been causing problems in slaughterhouses that were built to handle carcasses averaging less than 300 kg while today's average is just over 350 kg.

Processing meat

The next step-down in terms of total animal biomass moving along the meat chain is the conversion of carcasses to retail cuts, actions entailing considerable removal of bones and fat and meat trimmings, and further processing of raw meat ranging from simple grinding and mixing to prepare popular ground meats for frying and baking to elaborate aging, smoking, pickling and other ways of extending the durability of meat products. Carcass cutting yield, the ratio of meat to carcass weight, is affected by fatness and muscling (obviously, leaner and more muscular animals have higher meat yields), the degree of trim (the extent of fat left on meat cuts), the share of boneless cuts (the amount of edible meat remains the same but cutting yield is naturally higher with more bone-in cuts) and the leanness of ground meat (if all ground beef or pork from an animal is very fatty, the cutting yield will be higher).

Consequently, average cutting yields for beef and pork can span considerable ranges (Wulf 2012). For beef, they can go from as little as 58% for Holstein steer cut to produce closely trimmed boneless steaks and roasts and lean ground beef to as much as 65% for a heavily muscled, lean steer butchered into regularly trimmed bone-in cuts and fatty ground beef. For pork, they range from just 50% for a very fat hog cut into boneless, well-trimmed chops and roasts and lean ground meat to as much as 82% for a heavily muscled hog butchered into regular bone-in cuts and fatty ground meat. Multiplying the cutting yield by the carcass yields means that for US beef, as little as 29% and as much as 52% of live weight ends up as retail meat (typical range being 39–43%), and that the analogical range for pork is 37–62% (typical range 47–53%).

Modern meat-processing industry has changed everywhere as it had to scale up to meet the increasing, and shifting, demand for its products and as it became mechanized (and parts of it even fully automated), as well as restructured and concentrated, in order to cut its operating costs. New, larger meat-processing plants have been increasingly located in rural areas, and many old urban slaughterhouses were also relocated; this shift has been driven by the desire to reduce transportation costs of livestock to slaughterhouses, to ease environmental concerns and to weaken the power of labor unions, and it has created a new map of the meatpacking industry. Well-documented US trends capture all of these changes (Kandel and Parrado 2005). Increased amalgamation, integration and concentration of meat processing have resulted in numerous changes in ownership, and the regional concentration in the South is particularly evident for poultry processing plants (Muth et al. 2006).

In the US, the process has also created highly unstable labor force with rapid turnover rates and highly reliant on immigrant (particularly Hispanic) labor. To a large part, this is due to stressful and dangerous working conditions, with line speeds of up to 400 cattle processed per hour and with many workers having to perform on the order of 10,000 identical knife cuts a day. As a result, American meatpacking plants have very high rates of non-fatal occupational injury and illness among all private industries. Incidence rate of non-fatal occupational injuries (per 100 full-time employees working 40 hours a week for 50 weeks) for all industries is 3.6, 2.2 for mining but 6.0 for animal slaughtering and 5.3 for slaughtering and 5.8 for meat processing (BLS 2012).

Increased processing led to a substantial expansion of unskilled labor force in meatpacking plants where workers perform monotonous tasks (such as identical cuts with large knives). A major reason for a higher degree of processing was (as already noted) the decline in beef consumption that was, in mass terms, more than counterbalanced by the rise in demand for conveniently portioned chicken meat. During the 1950s, when beef dominated and chicken consumption was relatively low, less than 10% of chicken meat was retailed as cut-up products, but by the late 1990s that share was close to 90%. Large increases in US meat exports (those of beef and chicken rose nearly 60-fold between 1970 and 2000) were another major reason for increase in centralized meat processing.

In some traditional societies, killed animals were butchered rather indifferently (cleaver-hacked pieces and strips of Chinese pork hung from hooks for display in markets are a good example of this approach), but some pastoral societies used the distribution of carefully defined cuts as a way to express and enforce social status: recall, for example, the rigorously defined hierarchical habits of sharing described by Lokuruka (2006). Peasant societies of Europe had eventually developed minutely differentiated ways of butchering large animals that, inevitably, shared essential commonalities but that had also differed in national specifics.

Curious readers will find schematic butchering plans and illustration and description of specific meat cuts – that used to be made by independent butchers from carcasses hanging in their coolers and that are now increasingly mass-produced in large meatpacking plants – in many publications and comparisons that detail traditional American, British, Dutch, French or Italian varieties and explain the gradation from choice primal cuts to inferior pieces of meat and contrast principal differences in butchering cattle and hogs (Green 2005; Knox and Richmond 2012). Such traditional beef eaters as the English have some 35 distinct cuts, a fraction of more than 100 specific cuts available to beef-loving Koreans.

But besides the primary kind of meat processing (butchering of animals and preparation of conveniently packaged cuts), there is another large industry devoted to secondary processing that uses various meat cuts, fats, organs and skin to prepare a wide variety of fresh, cured, cooked, fermented and dried meat products. There is no single word describing these product groups. The French term *charcuterie* (cooked flesh) is traditionally restricted to processed pork-based products (bacon, ham, sausages, terrines, pates, confits), while Italian *salumi* and German *Fleischwaren* can contain meat from many species of mammals (including donkeys, wild boars or horses) as do sausages from Hungary or Asia.

Heinz and Hautzinger (2007) offer a simple but exhaustive division according to the methods of processing. Fresh processed meat products, dominated by hamburgers, chicken nuggets and kebabs, have emerged as new global fast-food favorites. Cured meats can be simply salted and air-dried (as is beef in Swiss *Bündnerfleisch* or pork in Spain's *jamón ibérico*) or raw-cooked (wieners, *mortadella*, *lyonnaise*, various meat loaves using one kind of meat or combinations). Pre-cooked products include such traditional favorites as liver sausage (*Leberwurst*), blood sausages and corned beef. The category of raw (dry) fermented sausages subsumes numerous kinds of European salami as well as some traditional Asian sausages.

Finally, dried meat comes usually in strips or flat pieces (biltong, beef jerkies) or as meat floss in Asia. The origins of most of these processing methods are obvious – to preserve meat in the absence of refrigeration – but their growing popularity in modern societies is based on three pillars of tradition (France without *pâté*, Toscana without *salumi*?), flavor (pepperoni is the most common topping on American pizza) and convenience (ready-to-eat or served after brief heating/broiling). Although processing extends shelf life of meat, most of the products must be still refrigerated and packaged in suitable plastics (films, extruded webs) that provide oxygen barrier and delay spoilage (Belcher 2006).

Before moving to appraisals of actual meat consumption, I have to take at least a brief look at the fate of all zoomass that is not part of dressed carcasses and that amounts to about 30% of live weight for pigs and up to 45% for cattle. Modern meat processing is a prime example of a frugal industry and minimized waste that turns virtually everything into something as it deals with three main categories of by-products, with edible offal, inedible offal and hides (Marti et al. 2011). Edible offal (organ or variety meats) makes up about 12% of live weight in cattle and 14% in pigs, and it includes numerous internal tissues – more commonly consumed livers, kidneys, hearts, tongues and tripe (part of bovine stomach), less commonly eaten brain and sweetbreads – as well as some organ bits cut off a carcass (feet, ears, tails), bloods and some marrow bones.

Traditionally, virtually all of these organs and cuts were readily consumed in the West and went into some classic regional signature meals (*tripa a la fiorentina*, Bavarian *saure Lunge* made with pork lungs and heart). Demand for edible offal began to decline as increased incomes make good meat cuts more affordable, and now most of it is domestically consumed in such products as sausages and pet food, and both the EU and the US are major exporters of these edible by-products. In 2010, the EU supplied 36% of all edible offal exports and the US share was about 28%, with Brazil and Canada ranking third and fourth. The largest destinations for the export of edible pork offal are China, Mexico, Japan and Russia; Chinese domestic prices are considerably higher than the US averages, particularly for pork kidneys, livers, hearts and tails. Japan has been the largest importer of US edible bovine offal (it bought more than half of the total in 2010), but recently Egypt became a large buyer of bovine livers and kidneys, with Mexico taking a majority of sweetbreads, tripe and tongue.

Cattle hides and pork skin come first to mind when thinking about inedible animal by-products, and despite a massive diffusion of plastics, they remain a valuable raw material processed by tanning into leather for shoes, bags, clothes, furniture and car interiors. But in mass terms, it is the residual mass produced by rendering (heat processing) of animal fats and bones that is most important: as much as 40% of bovine live weight ends up going through rendering plants where it is converted to high-protein feed for animals and pets and also to fertilizer. Finally, animal by-products are an essential source of hormones and enzymes derived from livestock glands (Aberle et al. 2001). That long medicinal list includes estrogens, insulin, parathyroid hormone, thyroid-stimulating hormone and testosterone; other life-saving compounds derived from animal tissues include antigens, antitoxins, serums, vaccines and blood product (fibrin, fibrinolysin and thrombin).

Consuming and wasting meat

Finding how much meat people actually consume might seem to be as easy as consulting appropriate tables in national statistical yearbooks. The electronic version of *Annuaire Statistique de France* has a spreadsheet going back to 1970 that is clearly entitled *Consommation moyenne de quelques produits alimentaires* and that lists (in kilograms per year per capita) consumption of beef and poultry – but not of pork. America's *Agricultural Statistics* contain a large table of "Per Capita Consumption of Major Food Commodities" that lists (in pounds per capita) annual consumption of beef, veal, lamb, mutton, pork, chicken and turkey. But detailed tables showing the breakdown by macro- and micronutrients in the same volume say "quantities available for consumption" (per capita per day) rather than simply consumption, and so do the US Department of Agriculture (USDA) spreadsheets for different kinds of meat that go back all the way to 1909. And *China Statistical Yearbook* displays "Per Capita Annual Purchases of Major Commodities" (in kilograms per capita for beef, pork, mutton and poultry) for both urban and rural households and for different levels of income.

Data for these major nations thus refer variously to consumption, to quantities available for consumption and to food purchased by households. And the best available global compilation of food statistics, FAO's food balance sheets, provide detailed disaggregations for energy, proteins and lipids for individual commodities that refer to food supply per capita. In fact, all of these differently labeled categories refer to an essentially identical concept, to food available at retail level (available supply, take-home food) and not to food that is actually consumed (eaten) at home, in institutional settings (hospitals, prisons, army barracks, office cafeterias) or bought for immediate consumption from fast-food outlets or enjoyed in restaurants. Unfortunately, too many users of national and international statistics do not pause even long enough to ascertain the actual meaning of presented categories and misleadingly refer to all of them as consumption.

Although the per capita figures from balance sheets are commonly mistaken for actual consumption rates, they do not tell us how much food people eat, only how much food is supplied. Obviously, not all supplied food (be it at wholesale or retail level) will reach consumers because of the losses that take place due to storage, transportation and retailing; not all food taken home will be eaten because of spoilage, food preparation and cooking losses and unconsumed leftovers. This means that reliable information about actual food intakes could come only from well-designed food surveys, and because prevailing habits and food preference keep changing, such investigations should be periodically repeated in order to uncover important consumption shifts.

Unfortunately, these are costly endeavors: representative food consumption surveys must examine a sufficiently large number of households, and they should be administered by trained personnel. That is why they have become increasingly rare, and even in the US, the National Health and Nutrition Examination Survey (NHANES), launched in 1971 and repeated in regular intervals, resorts to the least expensive method of dietary recall, with subjects providing a list of foods eaten during the previous day. This may be an expedient approach, but its accuracy cannot equal that of a supervised survey that involves actual measurement of consumed food quantities. Japan's National Health and Nutrition Survey (NHNS) – conducted by registered dieticians who weigh actual amounts of foods before and/or after preparation as well as food that is wasted and record the composition of meals that were eaten outside a home – is now the sole global example of this highly accurate but expensive approach.

The survey began in war-ravaged Tōkyō in 1945, and by 1948 it had nationwide coverage. Until 1963, it was done four times a year during three consecutive days; recently, it has been a single-day affair examining food intakes of more than 12,000 people in about 5,000 randomly selected households (Katanoda and Matsumura 2002). This globally unique and a highly reliable survey thus provides the only true nationwide account of actual food consumption, and it is also an excellent indicator of the share of supplied food that is wasted. In 2009, food balance sheet prepared by Japan's Ministry of Agriculture, Forestry and Fisheries credited the country with the gross per capita supply of 43.5 kg of meat (beef 9.3 kg, pork 18.3 kg, chicken 15.5 kg) or almost 120 g/day. In contrast, NHNS found actual daily meat intake at little below 80 g/day, a difference of roughly 35%.

NHNS also makes it possible to contrast population-wide means with gender- and age-specific rates (NIHN 2007). While the recent nationwide daily meat intakes were close to 80 g/day, women averaged less than 70 g and those older than 70 years of age less than 45 g, while men averaged more than 90 g and those in their 30s consumed just over 125 g of meat a day. American data from the NHANES indicate similar difference in meat eating. In 2001–2002, the average daily intakes of all meat for adult (20 years and over) males and females were, respectively, 198 and 120 g/capita (a 40% difference), with maxima for males in their 30s (231 g) and minima for women over 70 (86 g). These gender-based differences were particularly pronounced in the case of beef consumption (85 g/day for adult men, only 42 g/day for women); the gap was much smaller for eating chicken (60 vs. 46 g/day) and the rates were nearly identical (7 and 6 g/day) for eating turkey. In 2003–2004, total meat consumption for males was about 136 g/day and for females 91 g/day (33% difference) with similar age-based minima and maxima (Daniel et al. 2011).

Germany's *Nationale Verzehrsstudie II* – conducted between November 2005 and the end of 2006, two decades after the first such survey in West Germany – fell somewhere in between the American and the Japanese way of ascertaining actual food intakes (Max Rubner-Institut 2008). Interviews with some 20,000 participants from 500 townships covering the entire country and aged 14–80 years established their basic dietary and activity habits and health parameters; 24-hour recall was used as the primary tool for determining food consumption, and a random selection of participants completed two 4-day food-weighing protocols that established exact amounts of individual consumption.

Published data show expected gender-based differences with total per capita meat consumption (meat, sausages and all other processed products: *Fleisch, Wurstwaren, Fleischerzeugnisse*) averaging 103 g/day among men (and ranging from about 90 g/day in Schleswig-Holstein to nearly 120 g/day in Thüringen) and only half those rates (average 53 g, range of about 45–60 g/day) among women. German age-based rates peaked between 19 and 24 years for males (at 120 g/day) and were, as expected, lowest for women between 65 and 80 years of age (just 46 g/day). Rates for meat alone ranged from 35 to 50 g/day for men (with Rheinland-Pfalz and Hamburg in the lead) and from less than 20 to nearly 30 g/day for women. Thüringen's high male meat intake is mostly due to unusually high consumption of processed products: in excess of 80 g/day, it accounts for about 70% of total meat intake in that *land*.

In 1999, French national dietary survey (INCA 1) found mean meat intake (including all meat products) averaging 97.9 g/day, with men at 117 g and women at 82 g (Volatier et al. 2007). Less than a decade later, INCA 2 (conducted between late 2005 and April 2007 and based on seven-day written records of household food intake) reported a significantly higher national mean (118.8 g/day, 21% above the 1999 level) that was largely due to a higher rate of consumption among men: their meat, poultry, game, offal and meat product (*viande, volaille et gibier, abats, charcuterie*) intake averaged 145.4 g/day compared to 94.5 g/day for women (AFSSA 2009). Britain's 2001 National Diet & Nutrition Survey found that during the seven monitored days, men consumed 31% more meat (including all meat products, some of whose weight, such as in the country's favorite meat pies and pastries, was not all meat) than women, and that the peak consumption rate was, again, during the 30s (Henderson et al. 2002).

Similar gender-based differences are shown by other European dietary surveys – 20 of them were performed since the year 2000 in EU nations (Merten et al. 2011) – and if we had good data for actual meat consumption in low-income Asian, African and Latin American countries, they would undoubtedly show that a very similar partitioning – minima among older women, maxima among adolescent and adult males – has a virtually global validity. But there is no global commonality for income- and residence-based differentiation. Economic development has reduced, but far from eliminated, substantial differences in meat supply/consumption that existed in all societies in the early stages of modernization between the poorest and the richest households, that is, typically between higher-income urban and lower-income rural populations – but in affluent countries, the correlation has been reversed.

For example, during the latter half of the 19th century, average daily French meat ration was about 125 g/capita – but the rate for Paris was more than twice as high at about 270 g/capita (Barles 2007). Similarly, a century later the official Chinese statistics showed that in the year 2000, per capita meat purchases by urban (i.e., richer) households were nearly 80% higher than those of rural families (NBS 2001). But in affluent countries, the richer households now tend to consume less meat (but more fish) than the poorer ones. For example, in 2006 in Germany, the average daily rate for men in the lowest income quintile was 110 g/capita compared to 80 g/capita for men in the highest income quintile (Max Rubner-Institut 2008).

Meat waste is an important part of unacceptably high food losses at household and institutional level. A new study commissioned by the FAO shows that in mass terms, overall food losses now average about 100 kg/capita in Europe and North America, with most of this waste incurred during the consumption stage: specific wastage rates are at that level more than 10% for meat, about 25% for cereals and 20–30% for vegetables (Gustavsson et al. 2011). But when we consider the feed needed to produce meat, then it is clear that in terms of harvested phytomass meat waste in many affluent countries represents a greater loss of primary production than does the waste of cereals or vegetables because every unit of meat had required at least three units of feed whose cultivation had competed directly with the production of plant food.

Overall intensity of food waste is easy to appreciate just by comparing food per capita supply (from food balance sheets, now generally in excess of 3,000 kcal/day, for the EU at 3,700 kcal/day) and actual food intakes that are 30–40% lower. This range has been confirmed by Hall et al. (2009) whose model of metabolic and activity requirements put the most likely food intake of the US population between 1974 and 2003 at between 2,100 and 2,300 kcal/day – but those three decades saw the average food supply rising from about 3,000 to 3,700 kcal/day. Consequently, the US food waste rose from 28% of the retail supply in 1974 to about 40% 30 years later, while a UK survey found that British household food waste amounted to about 31% of purchased food (WRAP 2008). Sausages (some 440 million pieces a year), rashers of bacon (200 million) and meat-based, ready-to-eat and take-away meals (some 120 million) were among the items that were most often thrown away whole and unused.

Exceptionally, for the US we have had decades of official loss estimates – subtracting losses from primary (i.e., carcass) weight due to butchering and removal of bones and at retail and consumer level (including cooking losses and uneaten meat) – used by the USDA in order to calculate per capita meat availability (USDA 2012a). According to these data series, during the four decades between 1970 and 2010, actual US per capita beef consumption declined by 20% from just over 100 g/day to just over 80 g/day, pork eating remained basically unchanged at almost 35 g/day and poultry consumption had doubled to just over 50 g/day. I am not aware of any similar, consistent long-term series of loss estimates for any other country.

Japan has been the least wasteful of all affluent countries, and although it does not publish estimates of annual meat loss similar to the US series, its uniquely detailed food intake surveys make it possible to fix the difference between supply and intake with a high degree of accuracy: the Ministry of Agriculture, Forestry and Fisheries calculated the 2007 supply at 2,551 kcal/capita compared to the actual intake of 1,841 kcal/capita, a gap of about 28% (Smil and Kobayashi 2012). But because the supply is calculated for individual ingredients (pork, beef, chicken) and wasted food is estimated in terms of meals (chicken with rice, hamburger), it is impossible to make specific loss estimates. Household waste accounts for about half of the overall daily loss, retail stores and restaurants each added about 15% and the rest is wasted during food processing. Disaggregated figures show that in 2009, Japan's meat loss accounted for about 2.4% of total loss by weight, less than half the share (5.8%) of lost seafood: as expected, vegetables with nearly 50% waste and fruits with nearly 16% loss topped that list.

Making sense of meat statistics

This section would be superfluous if individual countries and international organizations expressed meat output and consumption in standard and readily comparable ways – or if they would at least always clearly state which one of several possible measures they are using rather than leaving us with undefined, or improperly defined, categories. As a result, the world of meat statistics abounds with assumed, undefined, poorly defined, misunderstood and misinterpreted categories that make both secular and international comparisons inaccurate and misleading. Even the organizations that should, and do, know better are not immune to this cavalier treatment. The very first table in USDA's annual *Livestock Slaughter Summary* lists "meat production" in its last column (USDA 2011) – while a quick study of the document shows that the mass refers to carcass weight of specific animals, not to any of the two categories (bone-in or boneless) of take-home (or retail level) meat.

But the USDA is one of a few agencies that in most of its statistics clearly distinguishes live weight of meat animals, their carcass weight and the weight of bone-in or boneless meat. Perhaps the most helpful are now a century-long series of "supply and disappearance" statistics that start in 1909: they tabulate supply (production, exports, imports, stock changes) and disappearance of meat (in total and per capita terms) measured in carcass, retail and boneless terms and are available in separate series for all red meat and all poultry (USDA 2012a). And starting in 1970, this series goes into an even greater detail as it traces meat losses from primary weight to retail, from retail/institutional level to consumers and loss due to cooking and uneaten food with the final per capita availability rate adjusted for loss at all levels (Muth et al. 2011).

In this series, assumption concerning the losses from primary to retail level had changed slightly from 30.2% during the 1970s to as much as 33.9% by 1994 but have stayed at 33.1% ever since 1996, while constants have been used for the losses between retail and consumer level and due to cooking and household waste, 4.3% for the first and 32% for the second loss. The latter rate has been questioned in a new study by Muth et al. (2011). This study used The Nielsen Company's Homescan® data for 2004 food purchases and the NHANES food consumption data for 2003–2004 to develop new consumer-level food loss estimates for all food categories, and its newly proposed estimate for meat and poultry losses differ substantially from the previous levels now used by the USDA. The largest change is for chicken (from 40% to just 15%); while new loss rate for beef is 20% compared to the old value of 32%, lamb went down to 20%

from 36%, pork loss has been reduced from 39% to 29%, but turkey loss has been slightly increased from 32% to 35%.

These reduced rates reflect the higher sales of trimmed, boneless, ready-to-cook, pre-cooked or fully cooked meat cuts and processed meats whose consumption results in lower losses at household level. Consequently, similar loss assumptions might be generally valid for other high-income countries with similar levels of meat consumption, but even the new, lower, rates are too high to be used for adjustments in low-income countries where higher cost of meat and traditional habits of eating a much larger share of carcass weight, particularly in the case of pigs and poultry, call for developing or estimating country-specific rates.

In the absence of such specific data, the best we can do to estimate actual meat consumption for most of the world countries is to apply approximate reductions to the FAO's readily available supply figures. Those per capita rates come from national food balance sheets that account for domestic production and the net trade of all important foodstuffs and that are usually constructed for a calendar year. In order to construct a fairly reliable estimate of national meat supply, it is necessary to know the total numbers of animals slaughtered and their carcass weight (or their live weight and apply appropriate adjustment ratios) and to account (if any) for exports and imports of live animals for slaughter (rather than for breeding) and chilled or frozen meat. Appropriate conversion factors are then applied to find the energy and nutrient values of meat and the totals are divided by a country's mid-year population and are usually expressed in per capita rates per day.

Countries with good output statistics prepare their own food balance sheets, and their data are used by the FAO in preparation of the annual global series of food supply; in order to make the series comprehensive, balance sheets for many low-income countries must be calculated in FAO's headquarters on the basis of the best (often very fragmentary) information. And then there are the member countries providing official but highly suspect information. Ever since China resumed its regular publication of economic statistics during the late 1970s, it has been clear that the quality of many of those numbers is very poor and that the official tendency toward exaggeration is common. Careful observers noted that the discrepancies between output and household consumption for meat (as well as for eggs and seafood) that were reported in the same statistical yearbook (the first number in the section on agriculture, the second one in the section on household consumption) were particularly large.

Moreover, as Lu (1998) pointed out, they had grown rapidly between the early 1980s and the late 1990s: by 1995, officially reported per capita

Table 3.1 Conversions for the weights of American pork in 2010.

To \ From	Live weight	Carcass weight	Retail weight	Kitchen weight	Eaten weight
Live weight		1.33	1.82	1.89	2.63
Carcass weight	0.75		1.37	1.43	2.08
Retail weight	0.55	0.73		1.04	1.47
Kitchen weight	0.53	0.70	0.96		1.41
Eaten weight	0.38	0.48	0.68	0.71	

Table 3.2 Conversions for the weights of American beef in 2010.

To \ From	Live weight	Carcass weight	Retail weight	Kitchen weight	Edible weight
Live weight		1.67	2.50	2.63	3.33
Carcass weight	0.60		1.49	1.56	2.04
Retail weight	0.40	0.67		1.04	1.30
Household weight	0.38	0.64	0.96		1.25
Edible weight	0.30	0.40	0.77	0.80	

output of meat was 2.6 times the rate of per capita consumption (and the differences for eggs and seafood were, respectively, 2.5- and 4-fold). Lu's deconstruction and reconstruction of official statistics concluded that the actual meat output in 1995 was overstated by 48%, and a year later, China's Statistical Bureau released its revised estimates for meat output in 1996, admitting that the original total was overreported by more than 20% (Colby and Greene 1998). China's agricultural statistics have undoubtedly improved during the past 15 years, but I would not even want to guess if the recent statistics have erred only by ±5% (an inevitable and tolerable error given the size and the nature of China's livestock sector) or if a systematic bias toward overestimation still produces totals at least 10% higher than the real output.

Perhaps the most useful information for any reader trying to make sense of assorted meat statistics is to refer to the conversion tables for different categories of pork (Table 3.1) and beef (Table 3.2) weights (expressed as fractions or multiples of live weight) that allow for adjustments from live (slaughter) weight to carcass weight to retail weight to weight at a consumer (household or kitchen) level to actually eaten meat. The last category is different from edible weight; depending on the meat category and cut, willingness to eat fat and offal and the degree of trimming and plate waste, edible weight will be slightly or significantly greater than the weight that is actually eaten. Approximate conclusions are that for pork and beef, about half of the carcass weight is boneless meat, and this means that for hogs no more than 40% of their live weight becomes boneless meat and only about 30% of cattle live weight ends up as boneless cuts.

These conversions are derived from live weights, slaughtering and butchering practices of beef cattle and hogs by American meatpacking industry and from the latest studies of American meat preparation and cooking as outlined in the Food Availability (per capita) Data System developed by USDA's Economic Research Service (USDA 2012a) and in a new study of food losses at consumer level by Muth et al. (2011). And here, for a comparison, is the sequence for a young Charolais bull indicative of typical French yields: fasting live weight of 740 kg is reduced to 429 kg of cold carcass (dressing share of 58%) that becomes (after removing 68 kg of bones, 30 kg of fat and 21 kg of waste) 306 kg of marketable beef (edible weight being 41% of live weight) almost equally divided between fast- and slow-cooking cuts (CIV 2012). Consequently, the American conversions are also applicable, with small adjustments, to beef and pork weight sequences in the EU, Australia, New Zealand and Japan, and to pigs produced by feeding in confinement around the world, and I will use them in Chapter 4 to calculate efficiencies of intensive meat production.

And although these approximate adjustments could be applied to FAO's meat availability figures (expressed in carcass weight) in order to get rough estimates of actually consumed quantities, it must be kept in mind that the use of such averages could lead to substantial errors when applied to low-weight cattle raised only by grazing (particularly to animals in arid subtropical region) as well as to unconfined or partially confined pigs that were reared, butchered and consumed in a traditional manner. Given the global popularity of chicken meat and the different ways of its consumption, I am showing conversion ranges rather than single figures for the broiler sequence (Table 3.3).

These calculations for broilers (at five to six weeks of age) are based on studies of carcasses composition for broilers in the US, Canada, Egypt and Brazil published by Perreault and Leeson (1992), Pelicano et al. (2005),

Table 3.3 Conversion table for the weights of broilers.

	Live weight	*Carcass weight*	*Edible weight*
Live weight		1.45–1.35	2.50–1.54
Carcass weight	0.69–0.74		1.81–1.14
Edible weight	0.40–0.65	0.58–0.88	

Shahin and Elazeem (2005) and Murawska et al. (2011). The leading broiler breeds include Anak, Cobb, Euribird, Hubbard, Ross and Sasso, and a complete management manual for Ross (Aviagen 2009) illustrates the complexity of modern poultry business. Depending on the breed, sex, diet and live weight at slaughter, eviscerated carcass is 69–74% of live weight (and even higher with the Asian-style carcass, with head and feet attached). Bone weight ranges from 12% to 20% of carcass weight, and hence the total edible weight (including all skin, fat and wing, back and neck meat) is as much as 88% of carcass and 65% of live weight. On the other hand, eating only skinless breast and legs (as these parts are commonly available in supermarkets) would amount to consuming only about 40% of broilers' live weight.

Mass production of meat in highly urbanized societies originates in large central animal feeding operations that can house tens of thousands of broilers (pictured here with water dispensers) and hundreds of thousands heads of of pigs or cattle. Photo by the USDA.

4

What It Takes to Produce Meat

Food cannot be produced without numerous environmental impacts – but their extent and intensity have been obviously multiplied by modern, mass-scale intensive crop cultivation and animal husbandry. During the past three decades, concerns about global environmental change have been dominated by worries about the pace and possible degree of global warming, but we should not forget that the evolution of pastoral and agricultural practices – converting natural ecosystems to cropland and grazing land, and intensified harvesting of the biosphere's primary productivity (photosynthetic production of crops) to be consumed directly by humans or to be used as feed for animals – has been the single largest cause of global land cover and land use change as well as a major consumer of fossil fuels and primary electricity, a leading source of aquatic and atmospheric pollution, and a growing emitter of greenhouse gases.

As long as per capita meat consumption remained relatively low (in traditional agricultures that fed most of the people, this was the case until the latter half of the 19th century), livestock's contribution to these local, regional and global environmental transformations remained modest. Indeed, a strong case can be made that the Old World's livestock was not only an integral part of traditional agroecosystems but, as one of its indispensable foundations, also a positive environmental factor.

Should We Eat Meat?: Evolution and Consequences of Modern Carnivory,
First Edition. Vaclav Smil.
© 2013 John Wiley & Sons, Ltd. Published 2013 by John Wiley & Sons, Ltd.

Large animals used primarily for draft power did not (as heavy modern machinery does) compact soil, they usually did not compete for the harvested phytomass with humans (they were energized largely by consuming crop residues or locally available grasses), they were a valuable source of organic fertilizer, and they provided small but important dietary contribution through their milk and meat production. Omnivorous pigs, raised by foraging or fed a wide range of edible wastes in semi-confinement, did not compete with humans either – and produced copious manure. And free-living poultry required only a minimum of supplementary feeding to produce eggs, some meat and the richest (highest nitrogen content) manure.

Livestock began to exert a greater environmental impact only during the latter half of the 19th century when the numbers of large animals increased due to rising demand for draft power and when higher disposable incomes made meat more affordable and led to rising consumption of beef and pork. The first demand began to decline after 1910 with the introduction of internal combustion engines, and in high-income countries draft animals were almost completely displaced by 1960, but rising consumption of meat has more than made up for this decrease of animal claims on harvested phytomass. Reliable US data allow an accurate comparison.

Just before 1920, cultivation of concentrate (mainly oats) and forage (mainly alfalfa and hay) feed for draft animals claimed about 25% of the country's farmland; half a century later, there were virtually no working animals on US farms, and an analogical switch took place in all rich European countries even more rapidly during the first two post-WW II decades. The latter half of the 20th century had also seen an unprecedented growth of pastoral populations in Asia and Africa and a very large expansion of commercial grazing in Latin America associated with extensive deforestation. And during the century's closing decades, agricultural mechanization had greatly reduced the numbers of draft animals throughout East and Southeast Asia, while meat demand increased substantially in virtually all of those countries (overwhelmingly vegetarian India being a major exception).

Traditional practices of raising livestock as an essential part of mixed farming enterprises have gradually intensified with rising numbers of animals per typical operation and with more intensive use of commercial feed mixtures. This, together with breeding and better veterinary care, brought a remarkable gain in the efficiency of feed conversion, particularly in pork and broiler production. The final stage of this process has been the emergence and growth of concentrated livestock operations where large numbers of cattle, pigs and poultry are raised in confinement on optimally formulated rations. These operations now have the highest feed conversion rates and the best meat productivities. Growing adoption of intensive livestock production in confinement has also raised many concerns about the treatment of animals.

By the 1980s, rising meat consumption became a new global environmental concern, and interdisciplinary studies also began to assess energy demands of livestock products and effects of that fuel and electricity consumption. Large-scale meat production has many obviously direct as well as many less obvious indirect impacts on land, water and the atmosphere. The most obvious effects include changes ranging from conversions of natural ecosystems (mainly deforestation driven by the quest for expanded grazing, and conversion of forests and grasslands to arable land used for cultivation of feed crops) to degradation of pastures and to offensive smells and polluted waters adjacent to some massive feedlots.

Water polluted by livestock wastes can be a problem even for a small operation (especially if nitrate-laden waste seeps into wells), and management of wastes and prevention of large-scale nitrate and phosphate releases is a challenge for all concentrated feeding operations. But in global terms, most of the aquatic and atmospheric impacts attributable to livestock do not arise from feeding large numbers of animals (be it in a dispersed fashion or in intensive confined operations) and dealing with their waste but from producing their feed. This is particularly true in all affluent countries where most crops are grown for feed. Before addressing all of these matters in a systematic and highly quantitative manner, I will explain the modalities of modern meat supply by reviewing the contribution of meat from pastures, continuing importance of mixed farming and increasing worldwide dominance of pig and poultry sectors by modern confined feeding that is based on optimal intake of balanced rations and that has resulted in some impressive gains in the efficiency of meat production.

But before I proceed, I should explain why my survey of environmental impacts of meat production will exclude the animals kept primarily for milk and eggs even though they are eventually slaughtered and their meat contributes to the overall supply of beef, mutton, goat and chicken meat. Given their large body mass, their longevity and their high feed requirements, the case for exclusion is made best with dairy cattle.

In traditional settings, cows would be usually bred and milked for at least eight, even ten, years, and even with a low output of 3,000 L of milk a year (the US mean in 2010 was above 9,000 L), they would produce about 65 GJ of milk compared to no more than 2 GJ of meat and fat (assuming carcass weight of about 300 kg), an equivalent of only about 3% of food energy in milk, while in protein terms the beef share would be only slightly higher at about 4% (output would be nearly 850 kg in milk and 35 kg in beef). Although the modern production changed all of these variables, the milk : meat ratios have remained nearly identical.

Modern dairy cows are typically slaughtered for meat (it goes mostly into ground beef and processed products: recently, almost 20% of US beef has come from dairy cows) at four years of age, and many are culled at younger ages. US statistics for 2010 and 2011 show annual slaughter of, respectively, 2.8 and 2.9 million dairy cows out of the total of 9 million of milk cows and heifers that have calved, implying an average offtake rate of about 30%. Even in just three years, a cow would produce (averaging at least 8,000 kg/year) 65 GJ of food energy compared to about 2.5 GJ in beef (assuming carcass weight of about 360 kg), meat and fat adding up to only about 4% of digestible energy in milk. And while laying hens, whose energy output in eggs greatly surpasses the energy content of their meat, are still killed and eaten in all low- and lower-income countries, in the US hen meat has been increasingly left out of the food chain.

During the first decade of the 21st century, the US government subsidized the serving of some 50,000 of hen meat in chicken patties and salads for school lunches, while Campbell Soup, formerly the meat's largest buyer, stopped using it in its products because of the poor quality of these exhausted birds (Morrison et al. 2009). Their fragility (due to osteoporosis), which has made their long-distance transportation difficult (but it is required due to a decreasing number of slaughterhouses that would accept the birds), has led to a search for on-farm methods of their killing and alternative disposal (Webster et al. 1996; Freeman et al. 2009) – and to new ways of using this meat as a partial substitute for pollock in the production of imitation crab sticks (Jin et al. 2011). As far as dairy cows and laying hens are concerned, their meat, if used at all, is clearly a marginal by-product of their primary functions, and environmental consequences of dairy and egg production must be thus attributed to those respective industries and not to beef and chicken meat supply. The same conclusion obviously applies to sheep and goat raised primarily for milk and eventually killed for meat.

Modern Meat Production: Practices and Trends

For millennia, all but a small fraction of domesticated meat was produced in only two fundamentally different settings, either by pastoralists exploiting grasslands (often with seasonal relocations to grazing areas involving long-distance transhumance) or by the settled practitioners of mixed farming who had closely integrated their animal husbandry with the cultivation of annual and perennial crops. Inevitably, there were some hybrid practices (such as the varieties of agropastoralism in parts of sub-Saharan Africa and seasonal exploitation of rich mountain pastures by farmers in the Alps), but throughout the Western world mixed farming remained the dominant mode of meat production until after WW II – and in global terms, it was still the single largest meat producer even at the beginning of the 21st century when, according to the FAO's estimates, it produced about 45% of all beef, pork and poultry meat (Steinfeld et al. 2006).

But that primacy was due to the fact that some 70% of all beef was still originating in mixed farming, while more than half of pork and some three-quarters of all poultry meat were produced in intensive, landless arrangements. This rapid post-1950 transition – based on a shift from locally produced and unprocessed or lightly processed feeds to commercial mixtures whose ingredients often originated not only in different regions but in different continents – had separated specialized production of pigs and broilers from cropping, and the average size of individual enterprises increased to match the rising demand of more affluent populations living mostly in large cities.

Within two generations, these concentrated, factory-like establishments relying on commercially formulated and widely distributed feeds became the norm for poultry production not only in all affluent nations but also for supplying the cities in modernizing countries of Latin America and Asia. Eventually, modern pig farming followed a similar concentration trend, and landless production now dominates all supply destined for large cities, but it still comes overwhelmingly from traditional mixed farming in many densely populated agroecosystems of China and Southeast Asia. And although the images of huge, open American feedlots are now seen as an iconic depiction of modern beef production, that way of intensive meat production is still mostly confined to North America and Australia, with most beef (about 70%) coming from mixed systems and from grazing (about 25%).

These new ways of large-scale meat production have not only severed the traditional close links between domesticated animals and the agroecosystems in which they were embedded (by receiving foraging opportunities and concentrate feed and by returning draft power, food-stuffs and manures), but they have also created many local and regional environmental problems whose scope and intensity will be reviewed in this chapter. Moreover, these new practices can be maintained only by relying on what has become a truly global industry of producing, trading, processing, formulating, enriching and distributing feedstuffs whose composition has been optimized in order to maximize typical feed conversion rates and to raise animals to their slaughter weights in fractions of time that was required in traditional settings. I will review the best available evidence of these efficiency rates and productivity changes and close the chapter by taking a brief look at perhaps the most emotional concern associated with the modern mass-scale meat production – how these establishments treat the animals.

Meat from pastures and mixed farming

Seré and Steinfeld (1996) classified livestock production systems into two principal categories, those limited solely to animals and those being a part of mixed farming arrangements. The two subdivisions within the first category are the modern intensive meat production in confinement and the traditional exploitation of grasslands, mostly in more marginal environments that are unsuitable for annual or permanent cropping, be it because of temperature, rainfall or terrain limitations. Pastoralists thus still occupy niches ranging from tropical grasslands inhabited by wild ungulates (Maasai) to high cold plateaus (Tibet, Andes) and to the Arctic lowlands of Siberia (reindeer herders of Chukotka).

In latitudinal progression, these pastoral ecosystems range from humid and subhumid tropics and subtropics with permanently high phytomass production and year-round grazing through arid and semi-arid, mostly subtropical, environments with pronounced seasonal vegetation pulse all the way to higher latitudes in temperate zones where animals must be confined during winter and provided with harvested feed until their return to seasonal pastures. At the beginning of the 21st century, a bit more than a quarter of all large ruminant livestock (cattle and camels) lived on grasslands, and the share was slightly higher (about a third) for small ruminants (sheep and goats).

In North America, the greatest concentrations of pastured livestock are in Alberta, Canada, and on the Great Plains, with the largest numbers in Texas, Oklahoma, Kansas, Colorado and Nebraska. Newborn calves remain on grasslands with their mothers for six to nine months, and after weaning they may remain on the same pasture or be shipped to other grazing grounds in order to gain 90–180 kg before they are sent to a feedlot. In Latin America, cattle numbers have been increasing on the grasslands of Colombia and Venezuela and above all in Brazil where cattle numbers rose from 50 million during the late 1950s to more than 90 million by 1975, 170 million by 2000 and 210 million by 2010.

Brazilian cattle ranching, causing large-scale destruction of natural vegetation *cerrado* and parts of the Amazon's tropical rain forest, has been driven by tax and credit incentives, with the large landholders (and most of the business is in large and medium properties) actually less interested in raising cattle than securing their land tenure (McAlpine et al. 2009). Less destructively, pampas of Argentina have produced beef for export since the late 19th century. In contrast to commercialized, and highly export-oriented, grassland-based beef production in the Americas, grazing throughout the Sahelian zone of Africa and in the more humid parts of the continent remains an overwhelmingly subsistence affair, with the new republic of South Sudan, Central African Republic and Kenya having particularly large concentrations of ruminant livestock.

Animals of low body weight and poor average productivity are kept as an important social asset, and the main nutritional contribution is their milk. Asia's pastoral societies have been in retreat throughout the central part of the continent, with the largest numbers of cattle and small ruminants remaining in southern Kazakhstan (even after a major decline during the post-Soviet era), Uzbekistan, Turkmenistan, Mongolia (down to 2.2 million heads of cattle in 2010 compared to 3.8 million in 2000) and Tibet. In contrast, Australia's beef industry has become highly commercialized, and it is strongly oriented toward export of both live animals and meat.

Slaughter weights of cattle raised on tropical and subtropical grasslands are lower than those of animals raised in mixed farming systems, and hence the proportion of beef produced by grazing is a bit lower than the global share of grazed cattle but that beef is strongly represented in intercontinental meat trade, with Brazil, Argentina and Australia in the lead. Europe is the only continent where cattle numbers have been steadily declining: between 1975 and 2010, they were almost halved from nearly 250 million heads to about 125 million.

Large patterns of global distribution and details of continental and national densities of pasture-based livestock can be seen in maps published by Wint and Robinson (2007) and Robinson et al. (2011); they used the classification of livestock production systems developed by Seré and Steinfeld (1996) to map the global distribution of principal farming systems. As all such global-scale efforts as this mapping have inevitable limitations, above all due to often substantial disagreements regarding the classification of land use and land cover, they give useful instant impressions of livestock distribution on worldwide and continental scale.

These maps can also be used to locate the extent and principal concentrations of mixed farming systems where livestock complements crop cultivations. Traditionally, the two most distinguishing features of these systems were their diversity of crops and animals: not only did they rely on cultivating a relatively wide range of crops but did so often in complex rotations or with interplanting, and they also kept several species of livestock. On European farms, it was common to keep horses, cattle, pigs and chicken, and geese, ducks, pigeons and rabbits were often present as well. In traditional Chinese agriculture, richer farms in the South had water buffaloes, pigs, ducks and carps.

Seré and Steinfeld (1996) divided mixed farming systems that combine often intensive cultivation of annual and perennial crops with livestock production into two categories on the basis of water supply to crops, into widely distributed rainfed and much more restricted irrigated agroecosystems. As is the case with grasslands, each of these categories has distinct attributes according to climatic regions, in temperate zones (and in climatically similar tropical highlands), in humid and subhumid tropics and subtropics, and in extensive arid and semi-arid tropics and subtropics. Mixed farming was the mainstay of traditional agriculture in all non-pastoral regions well into the 20th century, and before the advent of intensive feeding in confinement, mixed farming systems used to support virtually all pigs and poultry.

While horses were kept in large numbers in richer parts of Europe and in the US and Canada, large ruminants (cattle or water buffaloes) were the keystone animals of traditional integrated farming in all poorer regions, but meat was not even a secondary goal of their presence: draft power and manure were far more important, and the most common arrangement was to keep the animals in barns from which it was relatively easy to gather their manure. This began to change first, and slowly, in North America during the 1920s. Replacement of draft animals by tractors, rapidly diffusing in the US during the 1920s and 1930s, was the first driver separating animal and crop production, and Hilimire (2011) concluded that the price support enacted for designated agricultural commodities promoted specialization as it reduced the risk inherent in narrowing the production range. The third key enabling factor was the post-WW II diffusion of inexpensive synthetic nitrogen fertilizers (that obviated the need to keep animals as indispensable source of a key macronutrient), and by the 1960s America's mixed farming was in retreat.

Similarly, even in the most intensively cultivated parts of Europe, mixed farms were dominant until the early 1960s, particularly on poorer soils that were less suitable for arable cropping (Oomen et al. 1998). Given the high densities of cattle and water buffaloes in farming regions of Asia (in India, Southeast Asia and China), North America (dominated by the Corn Belt) and Brazil, it is not surprising that nearly three-quarters of the world's large ruminant livestock now live in the mixed systems, but at two-thirds the share for small ruminants (sheep and goats) is not much lower, with the largest numbers in India and Africa. In the year 2000, more than 30% of ice-free continental area was occupied by mixed crop–livestock systems that still dominate in Western countries, as well as densely populated parts of Asia, Africa and Latin America and that remain the leading way of global meat production: they accounted for more than two-fifths of the world's pork output and more than a quarter of all chicken meat production.

Most of these shares originate in mixed irrigated systems in Asia where the coexistence of crops, pigs and poultry has been the cornerstone of intensive farming for millennia. And while in the year 2000 60% of mixed farm operations was still under extensive management, Notenbaert (2009) found that most of the population practicing that kind of farming in Asia, Africa and Latin America was in the areas with high intensification potential, particularly in the southern part of Brazil, coastal Western Africa and large parts of India and China. China has been a prime example of this shift toward intensification of livestock production: by 2005, the proportion of pigs raised in intensive systems was about 75%, and the share of intensively produced poultry had reached 90% (Robinson et al. 2011).

As global urbanization proceeds (since 2008 more than half of the world's population has been living in the cities), it will only intensify the changes that have been reducing the importance of mixed farming during the past generation: traditional Asian and Latin American mixed farms will be pushed further away from cities whose suburban and peri-urban areas will be taken over by more profitable residential, commercial and industrial uses; consolidation into larger operations will be imperative in order to satisfy the scale of meat demand by megacities; and that trend will, almost inevitably, bring a higher degree of vertical integration. Pigs and broilers for Shanghai or Bangkok are already produced by mass-scale centralized feeding operations; before too long this will be the norm in all better-off parts of the modernizing world. That shift may increase meat supply and lower average prices, but, as I will show later in this chapter, it will make the lives of animals even less tolerable.

Confined animal feeding

During the 20th century, it became possible to locate modern, intensive, "landless" livestock-only facilities in virtually any environment, a feat made possible by large requisite inputs of energy used for sheltering the animals in heated and air-conditioned structures, mechanizing their feeding, watering and waste removal and, above all, supplying them with balanced feed rations. These operations began first (and on a still relatively small scale) with American broilers before WW II, and during the 1960s they began to spread to hog raising as well as to the last (feedlot) stage of beef production in the US and Canada, and they have been the subject of numerous reviews and analyses (Ball 1998; CME 2004; MacDonald and McBride 2009).

Landless pig and broiler operations now dominate on all continents except for Africa, with about 60% of all pork and roughly 75% of all poultry meat coming from confined feeding in 2010, and both shares will undoubtedly grow in the future. In contrast, and as already noted, beef production is different: in the affluent countries with the highest per capita beef output (the US, Canada, Australia), as well as in their modernizing counterparts (Argentina, Brazil), it continues to depend heavily on grazing; throughout Europe, it comes mostly from mixed farms where it depends on a combination of grazing and supplementary feeding; and in low-income countries with large cattle herds, it comes almost exclusively from grazing. As a result, the share of the worldwide beef production that originated in landless facilities was only about 6% at the beginning of the 21st century.

Landless feeding goes under generic (and inaccurate) labels of factory farming, industrial farming or industrial animal operations, or it is referred to by more accurate terms of intensive livestock operations or confined animal feeding operations (CAFOs). The last acronym is the best one, as confinement (resulting in high stocking densities) and feeding (resulting in maximized rates of weight gain) are their two defining attributes. In the US, animal feeding operations are defined as facilities that stable or otherwise confine animals for at least 45 days in any 12-month period and that hold no less than 1,000 animal units (live weight equivalent to 1,000 lb or about 450 kg).

This definition simply acknowledges the fact that any meat animal, be it poultry, monogastric pigs or small and large ruminants, can be reared to slaughter weight in this intensive manner. There are no universally valid conversions for livestock units (LU). Well-fed adult head of cattle, typical of European or North American herds, is the standard, a two-year-old heifer rates 0.8 LU, while large ruminants (cattle and water buffaloes) on other continents rate between less than 0.5 LU in Asia and 0.75 LU in

Latin America. Pigs at different stages of growth are equivalent to between 0.03 (piglets) and 0.3 LU (breeding sows go up to as much as 0.5 LU), a sheep or goat is equal to about 0.1 LU and a broiler rates less than 0.01 LU.

In practice, this means that more than 10,000 broilers or turkeys are now confined for their entire short life span in enclosed and, as need be, heated or air-conditioned structures, that more than 1,000 pigs are kept in a single partly or fully slatted house until they reach their slaughter weight of about 100 kg, and that the largest cattle feedlots that "finish" beef animals hold more than 50,000 heads. Environmentally based defini-tion of large American CAFOs is any facility designated as a point source of pollutants. In practice, this means operations with inventories of at least 1,000 heads of beef cattle (in a single lot) or 2,500 pigs weighing more than 25 kg or 30,000 broilers (these numbers imply housing in at least two separate structures).

The broiler industry was the first meat-producing activity to accomplish the transition from traditional small-scale operations to highly concentrated enterprises, with America's turkey producers not far behind. Post-WW II centralization raised the share of broilers originating on farms selling at least 100,000 birds a year from less than 30% in 1960 to about 95% by the year 2000, and by 2002 the average broiler farm sold more than 500,000 birds a year (MacDonald and McBride 2009). The production chain is made up of specialized breeder, hatchery and grow-out farms. Growers provide housing, heating and cooling, watering and labor under contract with processors who supply the chicks, feed, required medicines and often also the crews that collect and load the birds for trucking to the slaughter plants. A typical broiler house is a structure 12×150 m that can pro-duce (depending on the final bird weight) 115,000–135,000 birds a year; double-house operations are imperative for added economies of scale, and some growers have as many as 15–18 houses, shipping out more than two million birds from a single farm.

Pigs were traditionally fed any digestible phytomass (crop residues, tubers, food waste), and they were also allowed to forage rather than being kept in confinement; consequently, these animals took a year or more to reach slaughter weight of around 100 kg, and these practices are still common in many low-income countries where pigs mature slowly. In contrast, virtually all pork in affluent nations and the meat sold in large urban markets in low-income countries are produced by intensive feeding (with soybean meal supplying protein and grain corn being the most important source of carbohydrates) in total confinement, and the animals reach their slaughter weight in less than six months after weaning. Output fluctuations have been smoother by year-round farrowing, and distinct breeds are used for producing leaner meat with less trimmable fat and for bacon.

Centralization began to transform American hog production only during the 1960s: by that time, still no less than 60% of the country's grain farmers also kept hogs. Traditional operation was farrow-to-finish on a small scale, with animals housed at individual farms. In contrast, today's practice is to specialize in single stages of production: feeder pig producers raise pigs from birth to sell them for finishing; farrow-to-wean farms sell the pigs to wean-to-feeder operations; and feeder-to-finish farms grow the young animals to slaughter weight (Key and Bride 2007). Moreover, contract arrangements separating pig production from pig ownership became common, with growers receiving inputs and assistance from contractors who sell to packers or processors (the latter two can be contractors themselves).

By 2005, concentration progressed to such a degree that fewer than 2% of hog owners with operations including more than 5,000 animals accounted for 75% of the nation's pig inventory, while nearly 70% of owners with fewer than 100 animals held a mere 1% (Key and Bride 2007). The average number of market hogs removed per farm had quintupled between 1990 and 2005, and large finishing facilities now contain more than 1,000 pigs/house, and with six houses at a site the operation can produce 12,000 hogs a year. Another impressive way to indicate the concentration trend is to note that in 1965 there were more than one million hog farmers in the US; by the year 2000, their total fell below 100,000.

Long pregnancy and slow maturation of cattle make it impossible to replicate the degree of poultry or pig concentration during the early stages of large ruminant life cycle. That is why even in 2010 there were more than 700,000 cow-calf operations (and about 750,000 beef cow operations) in the US where the modern beef production has evolved as a sequence of three distinct operations (USDA 2010a; McBride and Mathews 2011): the cow-calf phase (six to eight months) that ends with weaning and relies almost solely on pasture; the stocker phase (six to ten months) that relies on summer pasture or another kind of roughage; and the finishing in feedlots (where steers, castrated males, surpass heifers, females that have not calved, because a small share of heifers must be kept for herd maintenance) that depends on commercial feeds designed to produce maximum daily gains with the lowest possible feed/gain ratios (Ball 1998; Jones 2001).

Consequently, at any given time even in the US, less than 15% of all beef cattle are found in feedlots, and those enormous beef feedlots whose pictures are often displayed as paragons of modern livestock industry are still uncommon even in the US and Canada. During the two decades between 1989 and 2009, the number of US cattle operations declined by 28%, while the average number of animals per operation rose by 36% to nearly 100 heads, and the operations with more than 500 heads accounted for nearly half of the total cattle inventory (USDA 2010a). While the average sizes of the first two operations have been increasing, impressive economies of scale prevail only at the finishing stage.

Grain feeding of livestock began after WW II in the US and Canada; a decade later, it started in Europe and the USSR; Japan became a large importer of feed during the 1970s. The first large commercial beef feedlot was set up in Texas in 1950, and until the mid-1960s small farmer feedlots with fewer than 1,000 animals were dominant as they marketed more than 60% of all cattle; since the early 1990s, most of the marketed cattle comes from lots with more than 30,000 heads, and by 2007 fewer than 300 largest feedlots shipped 60% of all animals (MacDonald and McBride 2009). Maximum numbers and the resulting feedlot rankings for America's largest CAFOs keep changing, but the largest facilities (now easily inspected on satellite images) ranked by their one-time maximum capacity are Five Rivers Cattle Feeding in Loveland, CO (820,000 heads), Cactus Feeders in Texas (520,000 heads) and Agri Beef in Boise, ID, with 350,000 heads (Lowe and Gereffi 2009).

An even better impression of what is meant by modern confined feeding is gained by looking at those operations in terms of actual stocking densities; areas of average feedlot pen or pig or birdhouse are per head. Often it seems that CAFOs can hardly squeeze in more animals without rendering them utterly immobile, overheating them from the tight compress of their bodies or, literally, suffocating them. Beef cattle are relatively best off, with Australia's minimum stocking densities being $9\,m^2$/standard cattle unit (SCU, an animal with live weight of $600\,kg$) but with most feedlots in Queensland allowing $12–20\,m^2$/SCU (Queensland Government 2010). Canadian stocking densities are $27\,m^2$/heifer on unpaved ground but only $8\,m^2$ on paved ground in lots without a shed and mere $4.5\,m^2$ on paved ground in a shed (Hurnik et al. 1991), while the American minima are between 10 and $14\,m^2$/animal, and actual densities in unpaved lots on the Great Plains are commonly between 27 and $36\,m^2$/head.

That is confinement with very modest mobility, while pigs and poultry are truly confined to spaces barely larger than the footprint of their bodies. The EU's Welfare of Livestock Regulations (EU 91/630) specify minima of just $0.4\,m^2$ for an animal weighing less than $50\,kg$, not even $1\,m^2$ ($0.93\,m^2$) for a 100-kg pig (just enough space to ensure that it does not touch another pig during hot summer days) and only $2\,m^2$ for sows. Canadian rules give even less space, just $0.76\,m^2/100\,kg$ pig on fully slatted floor (Connor et al. 1993), and in the US, an area as small as $0.65\,m^2$ could house a 100-kg pig on a slatted floor. Economics has been the relentless driver of this congestion. Powell et al. (1993) showed that all performance indicators were better (daily feed intake and average daily gain were higher and average feed conversion efficiency was lower) at the lower stocking density of $0.81\,m^2/animal$, and the economic benefits (total meat sold) were about 5% better at the minimum density of just $0.48\,m^2/pig$.

Identical consideration applies in the case of stocking densities of broilers as studies show the birds performing better when allowed more space. For example, Bilgilli and Hess (1995) found that going from 740 to $930\,cm^2$ (i.e., from a square with the side of 27 cm to one with the side of 30 cm, or 1 ft^2) per bird produced higher body weight, better feed conversion and lower mortality, and Thaxton et al. (2006) found that raising the rate from 25 to $55\,kg/m^2$ did not affect such stress indicators as cortisone, glucose and cholesterol but that the densities beyond $30\,kg/m^2$ had adversely affected growth as well as meat yield of the crowded birds. But, as Fairchild (2005) stressed, the broiler farmers cannot afford lower densities as they would not get expected meat production.

Remarkably, the US, the world's largest broiler producer, has no binding stocking restrictions but only National Chicken Council guidelines that provide just $560–650\,cm^2/bird$, the latter space being just marginally larger than the standard A4 page ($21.5 \times 28\,cm$). Only mature broiler breeders (3.6 kg live weight) get more space: Canadian regulations allow them $1,860\,cm^2$ on litter and $1,670\,cm^2$ (approximately a square of 41 cm) on slat or wire flooring (CARC 2003). The moves to ease the crowding have been risible: in 2010, UK's new Broiler Welfare Directive reduced the maximum stocking density by about 7%, from $42\,kg/m^2$ of live poultry to $39\,kg/m^2$. And even much larger turkeys get as little as $0.28\,m^2$ for light and $0.37\,m^2$ for heavy toms.

While there is no possibility to further increase these nearly extreme stocking densities (indeed, there has been some public and legislative pressure to lower them), there is still room for a higher share of American beef and pork production to come from CAFOs, and there are still some (albeit diminishing) economies of scale that could lead to further growth of both average and the largest confined feeding facilities. But these trends, as well as the diffusion of poultry- and pork-producing CAFOs in many modernizing countries where rising income creates higher demand for meat, depend not only on the cost and availability of feed but also on successful management of environmental impacts of such massive feeding facilities.

High animal densities may have lowered unit production costs but they have caused a number of concerns and complications: they have greatly increased the potential for the spread of infectious diseases (culls required to prevent further diffusion of viral pathogens have affected anywhere between tens of thousands to millions of birds); they have narrowed the genetic base of meat-producing livestock as similar genotypes of highly productive breeds and lines provide the bulk of the output; intensive feeding and rapid turnovers have resulted in enormous volumes of concentrated waste; and increased specialization has led to large-scale long-distance transportation of live animals – for example, in the US, more than a quarter of unfinished pigs are moved from one state to another (Hennessy et al. 2004) – and it has lengthened the average transport distance between producers and meat packers.

Animal feedstuffs

An overwhelming majority of feed consumed by livestock in 1900 was, as it has been for millennia, fed to draft animals, that is, to oxen, donkeys, water buffaloes and camels in poorer regions of the Old World and Latin America and to horses in richer parts of Europe and throughout North America (where mules were also important). In North America, production of this feed made an unprecedented claim on agricultural resources because abundant farmland made it easy to support large horse teams (on Midwestern farms and especially on the Great Plains and in California) and because draft horses were the domestic animals that received most often supplementary grain rations. At the beginning of the 20th century, working animals in the US were fed daily at least 4 kg of hay and 4 kg of grain (mostly oats or corn), while other horses received 2 kg of grain (Bailey 1908). When they reached their peak numbers during the second decade of the 20th century, America's draft horses and mules required (assuming average grain yields of 1.5 t/ha and hay yields of at least 3 t/ha) at least 30 Mha, or a quarter of the country's farmland, to secure their feed (Smil 1994).

Beef came from grass-fed animals, and while herds of cattle and flocks of sheep feeding solely on cellulosic phytomass on pastures were sometimes kept in densities above the sustainable capacity of the exploited grasslands and hence had considerable negative environmental impacts – no matter if poorly managed or carefully tended – they did not consume phytomass digestible by humans and did not pre-empt any potential food production; nor did pigs that were fed household waste or left to root for feed or brought to forage in forests, or the free-running poultry whose supplementary feeding consisted largely of small amounts of whole grain. In contrast, commercial success of modern meat production and the worldwide diffusion of CAFOs is predicated on reliable and relatively inexpensive supply of mixed (compound) feeds that are assembled entirely from domestic inputs or from ingredients imported from several continents and that can include a remarkable variety of phytomass and, not uncommonly, also processed animal matter.

Because of their balanced content of macro- and micronutrients, modern compound feeds are superior to traditional feeds (whole grains, milling residues or food processing wastes) that were generally too low in protein and whose composition often changed substantially as a surplus phytomass became available for feeding after harvests or as a by-product of crop processing of food preparation. Modern feeds should result in rapid weight gains while maintaining animals in good health; that the latter goal has been in some cases superseded by the former quest should be a matter of serious concern, and I will address it later in this chapter.

Compound feeds are produced for high-volume domestic and export sales by multinational companies, but they are also formulated in smaller, cooperative facilities for the use by individual farmers or prepared on site by the owners of large livestock feeding operations. Preparation of compound feeds (based on the formulation of balanced rations) began, as so many industrial innovations, during the closing decades of the 19th century. Two US companies, Cargill and Purina, pioneered the effort, and both of them are still very much active today: the first one as one of the world's leading agribusiness conglomerates, the other one absorbed in 2001 by Nestlé.

In 1928, Purina was the first company to sell pelletized chicken feed as well as first dry feed for weaned calves. Interwar advances in the understanding of animal feed requirements began the quest for optimization of rations for different animals and different stages of their growth, and after WW II, formulations of balanced rations by small-scale operations, or even by individual farmers, became easier with the introduction of premixes containing requisite micronutrients and antibiotics that could be easily blended with bulk carbohydrate and protein sources.

The best data on the global use of compound feed production are those assembled by the International Feed Industry Federation (IFIF) which notes that those statistics are difficult to gather but that the uncertainty is reduced by the fact that four big producers (the US, EU, China and Brazil) dominate the output with about 65% of all feed produced in 2010 (IFIF 2012). IFIF data show a rapid rise of compound feed output between 1975 and 1985 (from 290 to 440 Mt), followed by slower increases during the next 20 years, to 591 Mt in 2000 and to 626 Mt in 2005, with roughly 410 Mt of the latter total coming from the big four. By 2010, the output was 15% up to about 720 Mt, and in national terms in 2010, the US and China (both at 160 Mt) had a narrow lead over the EU (150 Mt), followed by Brazil (60 Mt), Mexico, Canada, Japan, Germany and Spain. In 2011, the total reached 873 Mt, but in per capita terms the post-2000 output has stabilized at about 96–97 kg/year (Gill 2006). Remarkably, despite a high level of industry's concentration, the IFIF estimated that about 300 Mt of feed was produced directly by on-farm mixing rather than in centralized commercial mills that are subject to regular auditing of their compounding practices in order to ensure the safety of the food chain.

Uncertainty of the global figure was illustrated by an estimate of the World Feed Panorama (2012) that put the Chinese output at less than 110 Mt rather than at 160 Mt. And an even bigger surprise came as a global survey commissioned by Alltech (one of the world's top ten animal health and nutrition companies and conducted by its regional managers in 132 countries) estimated the worldwide feed production at 873 Mt in 2011, with Asia (305 Mt) and China (175 Mt) as continental and national leaders (IFIF 2012). If the estimates are correct, China's per capita feed output would be above 125 kg/year, still less than half of the EU rate (about 300 kg/capita) and roughly a quarter of the US rate (close to 500 kg/capita) but an order of magnitude higher than in still largely vegetarian India.

And the shift toward Asia can also be seen by comparing production growth rates: while between 2000 and 2011 the annual US output grew by less than 10% and the EU total increased by less than 20%, China's feed production had tripled. The survey had also provided the breakdown according to the final use, with 44% of the total feed destined for poultry (the total including both meat and egg production) and 25% for ruminants (the total dominated worldwide by feeds for dairy cows and excluding a similar mass of dry matter forages and silages fed on farms). Most of the remaining 31% were fed to pigs, then horses, pets and fish (aquacultural feeds at nearly 30 Mt or less than 4% of the total, with the highest consumption in China and other countries of East Asia).

Compositions of feedstuffs have been analyzed in great detail, and many compilations list their energy content (gross energy and species-specific digestible and metabolizable energies), details of their protein quality (down to the presence of individual amino acids for protein-rich phytomass) and concentrations of micronutrients (NAS 1982). This knowledge makes it easy to admit all of these plants and by-products into modern feed chain in quantities and combinations designed to maximize feeding. They could be divided into four principal categories: concentrates, crop residues, processing by-products, and animal wastes and feedstuff of animal origin. The most important group are concentrate feeds with high energy density: they are the main source of carbohydrates and protein.

Corn (maize) has the highest metabolizable energy of all cereal grains (14.3 MJ/kg of dry matter compared to 13.8 MJ/kg for wheat, 13.5 MJ/kg for potatoes and 13.2 MJ/kg for barley), and hence it became an obvious choice in all intensive feeding operations. Obviously, in the US this nutritional preference was helped by the crop's traditional prominence and by the success in raising its average yields, thanks to the introduction of hybrid varieties (starting in the 1930s) combined with intensive fertilization, pesticide application and, in many growing regions, supplementary irrigation. The grain is now the key component of intensive feeding worldwide. Other main carbohydrates include barley, oats, rye, sorghum and wheat, and all of these grains are also secondary sources of protein, whose content ranges from a low of 10% in corn to a high of nearly 16% in durum wheat. Tubers are less important because of their lower energy density but are often fed in order to prevent their spoilage.

Feeding rations differ in specific composition but are fairly uniform as far as the proportions of main sources of carbohydrates and protein are concerned: broiler feed 60% grain, most of the rest is soybean meal (Al-Deseit 2009); the two feedstuffs now dominate pig feed, and grain is 65–85% of all feed for growing and finishing beef cattle. Protein is the key ingredient for meat production, and as most countries cannot produce enough of it, the global feed formulation relies on extensive international trade in protein-rich crops and concentrates. Protein supply is dominated by oilseed meals produced by extracting plant oils from soybeans, rapeseed, cottonseed, sunflower seeds, peanuts, palm and coconut oil (FAO 2004).

Those edible oils are used directly in cooking or are incorporated into solid spreads and mayonnaise, while the residual meals – containing about 20 MJ/kg and between more than 35% protein relative to dry matter (up to 52% in soybean meals, 38% in rapeseed meals) and available in the form of pressed cakes or pellets – have become the principal source of protein in compound feed. Soybeans dominate this supply as they now account for three-quarters of all protein going into compound feeds. A less expensive solvent extraction, introduced during the 1950s, replaced the traditional crushing: the oil yield is about 18% of raw seeds and the residual meal has about 10% moisture, 42–44% of protein and a bit of fat. Soybeans are an ancient legume grain of China, and Commodore Perry brought them from his expedition that reopened Japan to the world in

1853, but before WW II the beans were grown outside of East Asia only as a curiosity: they were grown on only a few thousand hectares in the US, but since the early 1970s, the legume has been sown every year on more than 20 Mha, and increasing yields (now approaching 3 t/ha in the US) pushed the overall harvest to more than 50 Mt of seeds by 1990 and to 90 Mt by 2010.

Brazilian production has grown even faster: in 1960, it was only 0.25 Mt, by 1990 it was almost 20 Mt, and it had more than doubled between 2000 and 2010 from almost 33 Mt to nearly 69 Mt. Argentina now comes third, followed by China. Soybeans are extensively traded (nearly 40% of the crop were sold abroad in 2011); the US still leads, but Brazil and Argentina are now providing more than half of global exports (Turzi 2012). China (with more than 40 Mt) is by far the largest importer followed by the EU, Japan and Mexico. The remainder of feed protein comes not only from other oilseed meals (most of them with between 35% and 40% protein) and grains but also from alfalfa (dehydrated meal has 20% protein) and from feedstuffs produced by processing virtually all available animal tissues and wastes (Sapkota et al. 2007). Fish meals (made from ground-up small species such as anchovies and sardines and from fish and crustacean processing by-products) have as much as 65% protein.

A large category of rendered animal protein feeds made from slaughterhouse waste includes meat meals and tankage meals made from dried and ground-up internal organs, feathers, hair and bones as well as from entire bodies of dead, dying, diseased or disabled animals. As already noted in Chapter 1, feeding rendered animal tissues to cattle has caused a worrisome (but fortunately limited) outbreak of bovine spongiform encephalopathy that added to an ongoing trend of declining beef consumption. And as nothing goes to waste, there are also feeds recycling animal waste, typically as dried cow dung, swine excrement and poultry litter from CAFOs. Numerous experiments have shown that feeding of wastes does not adversely affect either the quality or taste of meat, or milk composition or flavor – but safe utilization requires that wastes are free of drug residues and that they are properly processed (by heat treatment or ensiling) in order to destroy potential pathogens. The environmental benefits of this practice can be high because up to 60% of its dry matter can be digested by cattle.

Sapkota et al. (2007) reviewed feed ingredients legally used in the US. They include growth-boosting and prophylactic antibiotics (above all tetracyclines, macrolides and fluoroquinolones), by-products of drug manufacturing (spent mycelia and fermentation products), antihelmintics, organic acids to prevent mold growth, arsenicals, non-protein nitrogen (urea, ammonium sulfate), minerals (iron and magnesium salts and Cu, Co and Zn), vitamins (especially A, D, B12 and E), yeasts, flavors, enzymes (phytase, cellulase, lactase, etc. aiding in digestion), preservatives and even plastics (polyethylene as roughage replacement).

US beef has also been receiving growth promoting hormones (testosterone, estradiol and progesterone and their more stable synthetic analogs) through slow-release implants under the skin. This practice has been criticized as irresponsibly dangerous, but the compounds have been approved for use in 30 countries; their proper usage (very low doses compared to natural hormone levels and to levels found in other animal and plant foods) has been found safe by the FDA, FAO and WTO; while their most obvious economic consequence is cheaper beef, decreased demand for land and reduced emissions of greenhouse gases are the two positive environmental effects (Avery and Avery 2007; Thomsen 2011). Contaminants detected in feed include bacteria (most commonly *Salmonella* and *Escherichia*, many with antimicrobial resistance), mycotoxins (most often from contaminated cereal grains), dioxins and polychlorinated biphenyls (because these compounds are lipophilic, they bioaccumulate in fat tissue and enter the feed chain in rendered animal fats).

Substantial shares of phytomass used for feeding could be either directly consumed by people or the cropland on which they are grown could be used to cultivate food cultivars. This human-edible component of animal feeding is often used in calculations illustrating how many more people could be fed by a purely vegetarian diet or by diets with substantially reduced meat consumption, and because grains make up by far the largest share of feed concentrates, such exercises are often reduced to answering the question: what share of grains do we feed, nationally or globally, to livestock? A closer look shows how easy it is to make major errors in such quantifications and to end up with dubious conclusions.

To begin with, there are no accurate figures regarding the global totals of food grains, legumes, oilseeds, tubers, vegetables and fruits directly used for feeding. For some of these categories (especially for vegetables and fruits, given their low-energy densities and very low protein content and seasonal nature of their feeding), this hardly matters. Oilseeds are energy-dense, but only a very small share of their total harvest is fed directly to animals: the FAO estimated that only about 5% of all oilseeds and 3% of all soybeans go directly for feed. The rest is processed to extract edible oil, and the resulting meals become valuable concentrate feeds.

But it would be misleading to attribute this meal mass to the total of potentially edible phytomass. Edible plant oils, the most valuable contribution of these seeds to human nutrition, and traditionally the main reason for their cultivation, have already been extracted and have entered the food chain, and the expanded cultivation of oil for high-protein animal feed thus increases rather than reduces their supply. And the relatively small quantities of oilseeds that are used for food without processing – fresh or roasted sunflower, peanut or soybean seeds for snacking, and, above all, soybeans needed for traditional coagulated (bean curd) or fermented (mainly soy sauce and *miso*) products – are easily diverted before oil extraction, and their supply is in no way compromised by feed demand.

A relatively large share of roots and tubers is fed to animals (the FAO's global crop balance sheets put it at about 22%), but, as already noted, this phytomass has typically low energy density (also a very low protein and lipid content), and if it were not diverted to feeding, it would make a marginal addition to the global supply of plant food. Diversion of human-edible roots and tubers to animal feed should thus be seen as reducing potential plant food supply mainly in an indirect and partial sense as some of the land used to grow those feedstuffs could be planted to more nutritious crops. And so the concern about feeding edible crops to animals can be largely reduced to grains: not only are they by far the largest component of feed concentrates but they also have high energy density, and some of them have fairly high protein content.

But their count should not include milling residues, that is, typically 15% of grain mass lost during the milling of wheat and 30–35% lost during the milling of rice: globally their output is now nearly 300 Mt/year, they make excellent feedstuffs, and unless people will prefer to eat whole wheat flour and brown rice, they should not be counted as diversion of edible phytomass to animal feeding. The best available estimates indicate that while the mass of grain used for animal feed had more than doubled during the second half of the 20th century and increased by almost 10% between 2000 and 2010, the proportion of global grain harvest destined for livestock has not gone up since the 1960s.

The share rose from about 15% in 1900 to more than 20% by 1950 and to 35% in 1960, but afterward it has fluctuated within a fairly narrow range between 36% and 40%, rising with the overall harvest from less than 300 Mt in 1960 to almost 500 Mt in 1975, to just short of 700 Mt in the year 2000 and to about 750 Mt in 2005. I must stress that this aggregate refers to all animal feeds, including the concentrates used to produce milk and eggs and, the fastest expanding category of CAFO, freshwater and coastal aquaculture that consumes carbohydrate and plant protein feeds in addition to compound feedstuffs based on seafood (mainly small fish and cephalopods) zoomass. Meat animals claim the largest share of concentrate feed, but in some Western countries, dairy and egg production may account for as much as 40% of the total.

Using the FAO's crop balance sheet estimates shows that the share of total domestic supply of US grain used for all animal feeding (again, including milk, egg and fish production) declined from about 80% in 1960 to 70% by 1990 and to 60% by 2005. For comparison, highly reliable US statistics show feed grain (corn, sorghum, oats, barley and small quantities of wheat and rye but excluding milling by-products) rising to about 180 Mt by the year 2001 and then declining to almost 160 Mt in 2010; when expressed as a share of total domestic use (after excluding America's substantial grain exports), these quantities translated, respectively, to about 71% and 47%, with most of the difference explained by the rising post-2011 use of corn for the production of ethanol.

An overwhelming majority of this concentrate feed comes from coarse grains (corn, sorghum, barley, oats, rye and millet) whose global annual output has been recently about 1,400 Mt of which roughly 45% were fed to animals. All coarse grains are edible, but their cultivation for feed competes directly with food production only in those countries where they are staples (corn in Mexico, sorghum and millet in parts of Africa). Worldwide, most people do not prefer them as staples, and hence the competition between coarse grains and potential food production is, again, overwhelmingly an indirect (and also a partial) one as the land used to grow corn or oats could be planted to other, more favored, staples (but in most cases not to rice) or to other crops. I will quantify these considerations later in this chapter.

In contrast, other feeds used in meat production are either not digestible by humans (cellulosic residues and forages) or are unsuitable for human consumption (assorted processing by-products and food wastes). Minimum quantities of cellulosic roughages should be a part of healthy diets, supplementing energy and protein, for all ruminants. These feeds come most commonly in dry form (between 10% and 15% moisture) as crop residues (a diverse category of inedible phytomass dominated by cereal and leguminous straws and corn stover) and as fresh or dry forages. Cereal straws and corn stover have low energy density (digestible energy between 6 and 7 MJ/kg), and fresh alfalfa, grass hays and corn silage (75% moisture) have similar gross energy values, while the dry matter energy density of these feeds is around 18 MJ/kg.

Another large category of feedstuffs that could be used in formulating commercial rations are various food processing by-products. Their exploitation has been made easier by concentration of modern milling, extraction, canning and other food processing industries whose operation yields large quantities of waste phytomass that could be profitably sold for direct animal feeding or for blending into commercial feed mixtures. Cereal milling by-products include hulls and brans; fruit processing yields skins, cores, peels and pulp (the last two produced in particularly large volumes by the citrus juice industry); and almond processing yields hulls (roughly half of the total weight, and in California sold to dairy for feed) and shells (used as fuel).

Productivity efficiencies and changes

As already explained, meat production in traditional settings – where animals were fed no, or hardly any, phytomass that could be consumed by humans or whose cultivation pre-empted the growing of edible foods – did not create concerns about the efficiency of feed-to-food conversion as does the modern intensive meat production that depends on concentrate feeds that could be either also directly consumed by humans or that are grown on land that could be used to cultivate various food crops. This increasingly common situation leads to concerns about the extent of the competition between feed and food crops, particularly when considering the future growth of human population. That potential competition is highest for intensively produced pork and chicken meat as pigs and poultry are fed only concentrates in confinement: worldwide pork and poultry production now claims more than 75% of all grain- and oilseed-based feeds supplemented by small quantities of by-products and consumes more than 70% of all animal feed that is cultivated on arable land (Galloway et al. 2007).

In contrast, ruminants (cattle, sheep and goats) could produce meat as pure grazers, and usually they consume roughage – forages and crop residues that could not be eaten by humans and whose production does not pre-empt cultivation of food crops – even when they are managed in confinement, not allowed to graze and fed a diet dominated by concentrates. This basic dichotomy must be kept in mind when calculating and comparing feed conversion ratios of meat-producing animals: ruminants are inferior feed converters compared to non-ruminant species, but their inevitably high reliance on forages from non-arable land and on cellulosic crop residues of food crops that are not digestible by humans greatly moderates their claim on the contested feeds produced on land that could grow food crops.

But there is no difference between ruminants and non-ruminants when they consume concentrate feeds: such conversion will always entail substantial losses of potential food output. Even when feeding grain corn yielding 10 t/ha to broilers, the most efficient meat producers, much less food energy and less protein will be available in the resulting chicken meat than could be supplied by eating bread or pasta made from wheat yielding 8 t/ha grown in the same field. This is an inescapable consequence of energy and protein losses due to inevitable inefficiencies of heterotrophic metabolism and growth. Of course, this simple quantitative contrast ignores different qualities and palatabilities of meat and grain staples, but even so the consequences of the intervening loss remain obvious, and the highest possible efficiency in converting feed to meat is always desirable. At the same time, large specific differences in efficiencies of animal growth and metabolism limit the effect of interventions and improvements aimed at more efficient meat production.

Performance of food-producing animals and their metabolic needs have been studied in considerable detail, and with feed being the most expensive component of intensive animal production, a great deal of effort went into researching and formulating optimal diets, with specific rations adjusted to different growing stages (Subcommittee on Feed Intake 1987; Fuller 2004).

These measures reduced typical feeding periods and hence cut the time needed to produce marketable animals while producing more meat, or more meat with desirable characteristics, per head. The most detailed accounts of feeding requirements at different stages of animals' lives are summarized in a series of recommendations published by the US National Research Council (NRC); the latest revisions of these reports are NRC (1988) for swine, NRC (1994) for poultry and NRC (1996) for beef cattle.

The best long-term record of feeding efficiencies averaged at the national level has been published annually by the US Department of Agriculture (USDA), with data for pigs and beef cattle going as far back as 1910 and with the record for chicken starting in the mid-1930s. All feedstuffs (concentrates, crop residues and forages) are included in the total and are converted to a common denominator of corn feeding units (with gross energy of 15.3 MJ/kg) needed to produce a live weight unit of an animal. This has the advantage of capturing the entire feeding burden – but a disadvantage of lumping together concentrate feeds and roughages and thus obscuring a fundamental difference between those two kinds of feeds.

Pigs are the most efficient producers of mammalian meat (Miller et al. 1991; Whittemore 1993). This is mainly due to their low basal metabolism as they need nearly 40% less feed energy than would be expected for animals with their adult body mass (on the order of 100 kg). Remarkably, at the fastest point of their growth, pigs convert nearly two-thirds of all metabolized energy into new body mass, at a rate of at least 40% higher than for steer and higher than even for chickens whose highest conversion rate of metabolizable energy is between 50% and 60% (Miller et al. 1991). In addition, pigs also start reproducing early (at as little as four months of age), have short gestation time (pregnancy averages 114 days) and have large litters (ranging from 8 to 18 piglets).

Brief gestation results in birth weights that are a much smaller fraction of adult weight than is typical for mammals (just 1/300 for pigs compared to 1/20 for humans), and this is compensated for by very fast and very efficient early growth, with weaned piglets consuming only 1.25 units of feed for 1 unit of gain. As with any heterotroph, porcine conversion efficiency declines with age as maturing adults (greater than 70 kg of weight) need between three and four units of feed per unit of gain (NRC 1988; Miller et al. 1991). Slaughter weights of 90–130 kg are usually reached 100–160 days after weaning, and hence the entire period between birth and marketing has been compressed to less than six months, or to less than half the time for traditional pig.

Average life-long feed conversion ratios with optimized rations containing about 13.5 MJ/kg of metabolizable energy and about 15% protein range between 2.5 (for the best American practices) and 3.5 kg of feed/kg of live weight. Further adjustments must be made in order to account for feeding costs of the breeding stock, and after adding the effects of environmental stresses, diseases and premature mortality can double the overall feed/gain ratios. The USDA's series for pigs shows two decades of gradual decline from the feed/live weight efficiency of 6.7 in 1910 followed by stagnation and fluctuation between 5 and 6.5 for the rest of the 20th century, with the rates of 5.0–5.9 between 2000 and 2010. This lack of efficiency improvement is explained by the demand for less fatty meat: leaner animals preferred by modern consumers could not be produced as efficiently as older, more fatty, breeds because the conversion of metabolizable energy to protein in pork peaks at about 45%, while the conversion of feed energy to fat can have efficiencies in excess of 70%.

A rather broad range of feed conversion ratios for beef requires some methodical explanations. Large body mass (for American steer, the average slaughter weight is now about 570 kg), long gestation and lactation and, as noted earlier, a higher basal metabolic rate than expected for animals of their size translate into feed/live weight ratios of 7–9 for breeding females, rates that are at least 50% higher than for swine and nearly three times higher than for chickens. Adjustments for feed needed for reproduction, growth and maintenance of the entire herd raise these feed efficiency ratios to more than ten corn equivalent units. The USDA's nationwide series for all cattle and calves shows no clear trend since 1910 as it fluctuates between lows of about 9 and highs of 14, with rates between 12.1 and 13.0 during the first decade of the 21st century. Production of sheep and lambs is even more inefficient, with recent feed/ live weight gain ratios at 15–16 – but these animals are not usually produced by feeding concentrates in confinement.

I must reiterate that the recent feed conversion efficiencies showing an average ratio of about 12.5 kg of feed/kg of live weight of cattle and calves capture the complete cost of beef production (including life-long mainte-nance metabolism and feeding needs of sire and dam animals under all kinds of production systems) in equivalent feeding value of corn. This is a useful approach for aggregate accounting, but it confuses the comparison with feeding conversion ratios for non-ruminant animals, and it requires requisite adjustment before assessing its meaning for any potential com-petition with food production. Beef can be produced along a continuum of arrangements that can range from no concentrate feeds to intensive finishing in CAFOs.

At one extreme is purely grass-fed beef; in between is a combination of various length of grazing and feedlot finishing with animals spending typically anywhere between 120 and 170 days (many even more than 200 days) in a feedlot until they reach their slaughter weight at 450– 570 kg. Grazing animals do not compete for phytomass with humans, and hence their growth does not pre-empt the use of any farmland for growing food crops, and their feeding efficiency is a key variable in proper grassland management but not in any calculation of potentially contested phytomass. For animals raised entirely in confinement on commercial feeds, the efficiency ratio is the aggregate total of feed to final slaughter weight, and this practice represents the least energy-efficient beef production alterna-tive. Limited mobility and concentrated feed result in much faster daily gains, as much as 1–1.4 kg compared to less than 500 g for grazing cattle.

Adjusting these rates in order to reflect only the consumption of con-centrate feeds grown on arable land (and hence making the feed conver-sion rates for beef readily comparable with almost purely concentrate-based feeding of pigs and poultry) is not easy. Beef production in all affluent nations combines forage and concentrate feeding in a variety of practices that differ not only among countries but also among regions and that change with prices of concentrate feeds and with the abundance of grazing phytomass. More than 90% of cumulative feed consumption of cows, bulls and replacement heifers could be forages (grazing and harvested plants) supplemented by grains and protein feeds, while the reverse would be true for a steer after a year's stay in a feedlot. In the US in 2010, only about 15% of all US cattle (14 out of 94 million) were "on feed," that is, receiving rations of concentrate feed with only minimal inputs of roughage in order to produce high-quality beef.

While ruminants should normally consume minimal amounts of roughage in order to keep their digestive system working properly, it is possible, after a short period of gradual adjustment (a month may suffice), to rear them after weaning solely by feeding them with concentrate feeds whose largest components are plants whose cultivation may directly compete with the production of staple food crops. Slaughtering a 300-kg steer raised on pasture produces beef without concentrate feeding, while moving that steer to a feedlot and raising its weight to 530 kg by feeding it about 1.4 t of grains and protein feeds (and just 0.4 t of harvested forages) results in grain feed conversion ratio of about 6.1 (1,400 kg/230 kg feedlot gain).

Birds maintain higher body temperature than do mammals. Passerines (songbirds) have core body temperature as high as 40°C, and if their small size would not preclude their rearing for meat, their high metabolism would. Non-passerine species, including all common meat birds ranging from pigeons to turkeys, maintain their cores at 39.5°C compared to 37–38°C in mammals, and this explains most of the difference in specific metabolic rates of these two vertebrate classes. But a rapid growth rate means that no other terrestrial domestic meat animal reaches slaughter weight faster than chicken as the average time needed to produce an American broiler was reduced from 72 days in 1960 to 48 days in 1995 as the average slaughter weight rose from 1.8 to 2.2 kg (Rinehart 1996; Boyd 2003; Havenstein 2006). Cumulative feed consumption of lighter birds fed balanced rations with 13.5 MJ/kg of metabolizable energy and 21% of protein and killed at five weeks of age will be on the order of 2.5 kg of grain, and feed/gain ratios may be as low as 1.5–1.8; for heavier broilers, the rates will rise to 1.8–2.0 (NRC 1994). About 10% should be added in order to account for the metabolism of breeder hens and cockerels and those birds that do not reach slaughter age.

In the mid-1930s, the average feeding efficiency ratio for broilers was the same as for pigs at about 5. Continuous improvements had halved this mean by the mid-1980s and lowered it to less than 2 by 2010 (so far, 2008 was the best year with the ratio at 1.75), a trend that contrasts with stagnating conversion efficiencies for beef and pork. No other domesticated bird is such an efficient feed converter as chicken: the rates for ducks (feed/live weight at slaughter) range from 2.5 to 2.9, and for much larger and older turkeys they have been recently just over 5.

The USDA's feed/live weight gain ratios, ranging from less than 2 to more than 12 for the three most common meat species, are perhaps the most often cited figures of feeding efficiency and are of the highest practical interest to producers interested in minimizing the amount of feed needed to attain typical slaughter weight. But, obviously, they are not the best indicators of energy and protein costs of actually consumed meat: as already explained, the share of edible beef may be less than 40% of live weight, and the difference between live and edible weight will be lowest in the case of traditional Chinese chickens (where only feathers, bones and beak are left uneaten). And, to repeat, they do not allow for a meaningful comparison between beef cattle and non-ruminant animals.

That is why I will recalculate all feeding ratios in terms of feed to edible product, provide the conversion rates both for energy and protein, and show separate energy and protein conversion ratios for concentrate feed. For energy, I will use gross feed input and the average energy content of edible tissues, and I will express protein conversion efficiencies as direct ratios (units of feed protein needed to produce a unit of edible protein) as well as in terms of gross feed energy per unit of food protein. And I will also recalculate the needs only in terms of concentrate feed whose cultivation puts livestock in potentially direct competition with food crops.

Table 4.1 summarizes rounded values needed to construct different efficiency ratios for total (concentrates and roughages) feed consumption, and it shows that in mass terms (kg of feed/kg of edible meat and associated

Table 4.1 Typical energy and protein conversion efficiencies of feed inputs to edible meat in the US.

	Chicken	Pork	Beef
Feed (kg/kg LW)*	2	5	10
Edible weight (% LW)[†]	60	53	40
Feed (kg/kg EW)	3.3	9.4	25
Energy content of feed (MJ/kg)[‡]	15	15	15
Energy content of meat (MJ/kg EW)[§]	7.5	13.0	13.4
Energy conversion efficiency (%)	15	9.2	3.6
Protein content of feed (%)[¶]	20	15	15
Protein content of meat (% of EW)[§]	20	14	15
Protein conversion efficiency (%)	30	10	4
Feed energy per unit of protein (MJ/g)	2.5	10	25

* Rounded values of recent performance from USDA.
[†] See Tables 3.1 and Table 3.2.
[‡] Gross energy content; metabolizable energy content is about 13.5 MJ/kg.
[§] Averages for edible part of carcass.
[¶] Typical means of protein content of concentrate feeds dominated by corn and soybeans.

fat) the recent American feeding efficiencies have averaged about 25 for beef, more than 9 for pork and more than 3 for chicken. These rates translate into energy efficiency of less than 4% for beef, nearly 10% for pork and 15% for chicken. Protein conversion efficiencies are, respectively, 4%, 10% and 30%, and while about 25 MJ of feed energy are needed to produce a gram of beef protein, the rates are only about 10 MJ/g for pork and 2.5 MJ/g for chicken.

For comparison, a detailed evaluation of four beef-producing sequences in the UK yielded feed conversion ratios as high as 27.5 kg of dry matter feed for a kg of bone-in carcass upland beef (spring calving, grass finishing over 80 weeks) and as low as 7.8 kg for "cereal" beef, that is, for calves finished in feedlot in 54 weeks (Wilkinson 2011). But in the first instances, less than 10% of all dry matter feed came from concentrates, while the share of concentrates was 95% in the latter instance where the cumulative feed intake of calf reared in confinement and slaughtered after a year when the animal's live weight will be about 540 kg was about 2,300 kg of concentrates and 90 kg of silage or hay. Koknaroglu et al. (2007) calculated that Iowa cattle in a feedlot (gaining 1.31 kg/day by ingesting 9.16 kg of dry matter) needed 6.26 kg of feed/kg of gain, while Lesschen et al. (2011) put the mean conversion values in the EU-27 means calculated at 19.8 for beef, 4.1 for pork and 3.3 for poultry (1.2 for cow's milk).

In order to express these rates as energy and protein conversion efficiencies, it is necessary to choose average energy and protein content of all edible meat for a particular species. In my calculations, I use approximate rates of 7.5 MJ/kg for chicken, and 13 MJ/kg for beef and for pork, and I assume the following shares of protein in edible portions: 20% for broilers, 15% for beef and for pork. Energy conversion efficiencies (energy content of feed/energy content of edible meat) are less than 3% for the entire US beef herd, 4–5% for combinations of grazing and feedlot finishing, and 6–8% for the gains during the time of intensive feedlot feeding; energy efficiencies are mostly between 8–10% for pork and 10–12% (and up to 15%) for broilers. Protein conversion efficiencies (protein in feed/protein in edible meat) are no more than 4% for the entire beef herd, as much as 10% for pork and up to 33% for chicken (Table 4.1).

Treatment of animals

This section, concentrating on pain, suffering, death and moral rights of domesticated animals, has to be very different from the rest of this factual, highly quantitative chapter teeming with numbers, rates and comparison. And making it just a section is an exercise in ruthless exclusion as this unavoidable and emotional topic (that is too often avoided and too rarely treated in balanced ways) consists of many concerns, admits arguments from many perspectives and has been covered (exhaustively and from many angles) by writers ranging from professional philosophers and ethicists to anti-meat crusaders and vegan activists. Perhaps the most expedient way to address this complex matter is to contrast some extreme positions with everyday reality of animal slaughter and suffering.

Aversion to killing animals is an ancient notion, and it has been one of the guiding tenets of Buddhism (Chapple 1993). The concept of *ahimsa* is defined by Vyasa's commentary on *Yoga Sutras* as "the absence of injuriousness (*anabhidroha*) toward all living things (*sarvabhuta*) in all respects (*sarvatha*) and for all times (*sarvada*)." No killing of living creatures ("sentient beings" in the standard Buddhist parlance) is permitted, although in practice there has been much inconsistency as even some traditionally strict Buddhist societies made large (Japanese Buddhism never proscribed eating fish and marine mammals) and small exceptions (allowing meat eating in sickness), and the less strict ones had found creative solutions to circumvent that absolute ban: as already noted, Japanese eating of wild boar was approved once labeled a mountain whale (*yama kujira*).

Modern expressions of these ancient concerns are composed of (to some extent commingled) strands of theoretical treatises on the rights of animals, often vigorous (and sometimes even violent) protest actions by national and international organizations defending all animals against mistreatment and slaughter, and (in a non-radical way) practical proposal of groups aiming at reducing all avoidable suffering. In order to minimize the distinction between our species and other mammals, many advocates – determined to expose human "speciesism" – stress that theirs is a fight for the rights of non-human animals subject to a highly amoral and indefensible behavior of human animals.

Singer's *Animal Liberation* remains one of the canonical texts of modern movement against animal slaughter (Singer 1975). That it has generated such a strong reaction is not surprising: facing ethical problems raised by his arguments can be uncomfortable (Dardenne 2010). My concern, illustrated by just two quotes from the book, is that Singer mixes provocative but necessary arguments with conclusions carried *ad absurdum*. He starts with a morally persuasive argument (Singer 1975, 8):

> If a being suffers there can be no moral justification for refusing to take that suffering into consideration. No matter what the nature of the being, the principle of equality requires that its suffering be counted equally with the like suffering – insofar as rough comparisons can be made – of any other

being. So the limit of sentience is the only defensible boundary of concern for the interests of others. To mark this boundary by some other characteristic like intelligence or rationality would be to mark it in an arbitrary manner.

But the logic he then follows would make feeding the world impossible, not because humans could not eat any meat but because they could not even kill pests that could ravage their crops: "No consideration at all is given to the interests of the 'pests' – the very word 'pest' seems to exclude any concern for the animals themselves. But the classification 'pest' is our own, and a rabbit that is a pest is as capable of suffering, and as deserving of consideration, as a white rabbit who is a beloved companion animal" (Singer 1975). Following this advice would mean surrendering a large share of our food harvest to heterotrophs; how large would depend on where this Singerian consideration would stop. Obviously, he would let all mammals feed on farm crops, but would he include all birds and would he exclude insects?

Singer's work was followed by many, and sometimes lengthy, examinations including those of DeGrazia (1996), Cavalieri (2003), Scruton (2006), Smith (2010), Waldau (2011) and Scully (2011). Some of these writings argue that non-human animals deserve human rights, a demand that could not be easily put to practice given the very nature of food webs where everybody gets eventually eaten; but perhaps wild animals are less deserving of human rights than the domesticated ones. Animal rights activities benefited from the Internet as all those numerous national and international animal rights organizations advocate, as all of them have active websites. Their arguments and messages span a wide spectrum from Buddhist-like promotion of veganism to calls for violence, and they include celebrity endorsements as well as practical proposals by non-ideological groups whose main interest is to improve the lot of domestic animals.

America's Animal Welfare Institute, the EU Coalition for Farm Animals, Compassion in World Farming Trust and the British Royal Society for the Prevention of Cruelty to Animals and the World Society for the Protection of Animals are prominent in the latter category, while PETA (People for Ethical Treatment of Animals), set up in 1980, relies on celebrity appeals and protests for disseminating its leading motto that "animals are not ours to eat, wear, experiment on, or use for entertainment" (PETA 2012). The organization is uncompromisingly vegan, does not tolerate even the use of wool and has also been involved in rescuing pets (most of whom are subsequently killed); it is against the use of animals in testing of commercial products.

The Animal Liberation Front is perhaps the leading proponent of militant action: anybody mistreating or killing an animal is a terrorist, and animal liberation is to be achieved "by any means necessary" (ALF 2012). And there are numerous vegan groups (The Vegan Society, Vegan Action, Veganpeace, veganism.com) and vegetarian associations (besides all animal rights organizations, leading proponents congregate at The Vegetarian Resource Group). All of them ignore the realities of human evolution and humanity's undeniable omnivory, extol the virtues of grains, legumes, vegetables and fruits, some going even as far as advocating eating only raw plant tissues or subsisting solely on fruits.

Philosophical or militant, theoretical or practically oriented, advocates of animal rights do raise some fundamental concerns. I have already noted many instances of poor treatment and unnecessary pain inflicted on animals during their (often unnaturally rapid) growth, transport to slaughter and actual killing, and many more are detailed in books that have focused on the negative aspects of modern meat production, including those by Mason and Singer (1990), Lovenheim (2002) and Scully (2011). Some of these abuses can be, as I have also already noted, obviated or alleviated by simple measures. Unfortunately, as shown by the deplorable treatment of animals in some slaughterhouses, a greater human–livestock intimacy is no deterrent to cruelty (Purcell 2011), and rules and laws are required to minimize abuses.

As much as I deplore any accidental or thoughtless mistreatment of domesticated animals during transport and just prior to slaughter, my principal concern is about those deliberate choices that make the animals suffer, often on truly massive scales, for days, weeks or months. If a single example of this poor treatment is to be chosen according to the quantity of suffering, then none rivals the way of raising billions of modern broilers. All of America's leading broiler producers (Tyson, Pilgrim's Pride, Perdue) follow only the voluntary guidelines of the National Chicken Council whose latest update, approved in January 2010 (NCC 2010), allows more space than the previous recommendations but still fails to meet acceptable conditions. Their first rule of management practices says that "Birds are allowed to roam freely throughout the growing area," but half a dozen lines below that statement, it specifies a minimum of 1.3 ft²/bird in the rearing house, an area smaller than two sheets of letter (A4) paper, hardly a space suitable for free roaming.

Dimness that prevails in broiler houses (0.5-ft candle or 5.38 lux) is merely an equivalent of a moonlit night or roughly just 1/100 of light delivered by a typical fluorescent light in modern kitchens. As distressing as living in high density in dimly lit spaces on an accumulated layer of excrement may be – the first condition prevents a normal range of activities, the second one destroys normal circadian rhythm, the last one damages feet and burns skin – that dismal combination has not been the worst part of the deliberate treatment of broilers: that dubious title belongs to the selection for excessive growth of their breast meat that induces suffering that is both grotesque and deadly (RSPCA 2002; Turner et al. 2005; AWI 2010). Rapid growth of breast muscle transfers the center of the bird's gravity forward, impairs the bird's already restricted movement and creates serious leg problems, and it also strains its cardiovascular system. Low price of broiler meat and its increasing affordability for poorer segments of modern society could be the only defenses of this reprehensible practice.

Europe has been ahead of the US in outlawing the most offensive practices. Matheny and Leahy (2007) made a detailed comparison of farm-animal legislation in the US and Europe where several common, and some national, standards allow more space and ban a number of practices that are still legal in US welfare (including gestation crates for sows and beak trimming of broilers). What is perhaps most unacceptable is that so much more could be done to limit the suffering of animals without any drastic increases in the cost of meat production: there is simply no acceptable excuse for claiming that the CAFO practices evolved during the last two generations represent optimum solutions to be followed, unchanged, in coming decades.

Meat: An Environmentally Expensive Food

Pastoral meat production requires large areas of grazing land, in poorer grasslands no less than 1 ha for a head of cattle, and this makes pastures the world's most extensive anthropogenic ecosystems marked by reduced biodiversity, frequent overgrazing, increased soil erosion and desertification and vulnerable to climate change. Modern mass production of meat in confinement is predicated on intensive cultivation of feedstuffs, on adequate watering and on the provision of suitable shelter (this may require seasonal heating or air-conditioning). And because high animal density in CAFOs could result in rapid diffusion of infections, mass production of meat has also been predicated on prophylactic applications of antibiotics, some of which also act as growth promoters (Landers et al. 2012), and on a constant removal and effective management of copious wastes. An insufficient or negligent provision of some of them will rapidly compromise or doom the enterprise. Not surprisingly, demands of these landless livestock operations translate into a multitude of environmental impacts.

As already explained in this chapter, large mammals are relatively inefficient convertors of feed to meat, and hence they require relatively large amounts of feed. Given the high levels of meat intakes in all Western nations, this means that those countries now devote most (or no less than half) of their cultivated land to feed crops rather than to cultivation of food plants. Most of the consequences of modern intensive crop cultivation – including the reduction of agroecosystemic biodiversity due to reduced crop rotation and large-scale cultivation of extensive monocultures, high rates of erosion accompanying the early growth of row crops, considerable losses of applied nitrogenous fertilizers and deleterious impact of agrochemicals – should thus be attributed to meat production.

Low feed conversion ratios also mean that the overall energy cost of meat production must be very high because feed crop cultivation is predicated on considerable direct (in fuels and electricity) and indirect energy inputs (above all, energies embodied in fertilizers and other agrochemicals). To this must be added direct and indirect energy uses required to maintain suitable living conditions (particularly for air-conditioning of large broiler houses in summer) and to remove and to treat wastes. Inevitably, energy costs of meat must be multiples of energy cost of plant foods, or of aquacultured herbivorous aquatic species.

Waste generated by modern animal husbandry has become a major source of not just local but also regional environmental pollution. Volatilization of ammonia is the source of objectionable odors from large-scale operations, particularly dairy farms and piggeries; the gas also contributes to both eutrophication and acidification of terrestrial ecosystems. Leaching of nitrates and the resulting contamination and eutrophication of waters has been given perhaps most of the attention, but accumulation of phosphorus and heavy metals – copper, zinc and cadmium originating in fertilizers used to grow feed crops and in compounds added to animal diet – is also a serious problem.

Aquatic impacts are often the most evident and most objectionable side effects of large-scale meat production, and with larger concentrations of animals spread over larger areas, their effects have expanded from what used to be highly localized impacts to regional environmental degradation whose effects (particularly eutrophication of coastal waters) can be sometimes felt hundreds of kilometers from their sources. Concerns about atmospheric impacts of large-scale meat production used to be limited to ammonia emissions and concentrations inside and in the immediate vicinity of barns or birdhouses as well as downwind from such large regional concentrations of animals as the US Corn Belt or the southern part of the Netherlands.

Those concerns have not disappeared, but the worries about potentially rapid global warming have shifted the focus to the emissions of three greenhouse gases associated with livestock: carbon dioxide, mainly from the conversion of forests to grasslands and from the cultivation of feed crops; methane from enteric fermentation of ruminant animals; and nitrous oxide from fertilizers used to grow animal feed and from denitrification and nitrification of animal wastes. Precise apportioning is impossible, but there is no doubt that CH_4 and N_2O emissions attributable to meat production account for single largest shares of global emissions of those two leading greenhouse gases. While these disparate environmental impacts cannot be assessed in an aggregate fashion (unlike, e.g., various greenhouse gases whose impact can be compared by converting their effect to a common denominator equivalent to radiative forcing of CO_2), there can be no doubt that their combined effect amounts to a uniquely widespread and worrisome impact.

Its individual components have received an increasing amount of research attention, and its totality was examined in the FAO's *Livestock's Long Shadow* (Steinfeld et al. 2006). A more recent assessment by de Vries and de Boer (2010) evaluated and compared 16 life-cycle analyses quantifying environmental impacts for livestock products and ended up with predictable results: in overall terms, beef is the most demanding meat. Production of 1 kg of beef needs 27–49 m² of land compared to about 9–12 m²/kg for pork and 8–10 m²/kg for chicken (analogical rates are 144–258, 47–64 and 42–52 m²/kg of protein), and its global warming potential (GWP) is equal to 14–32 kg of CO_2 compared to about 4–10 kg for pork and 4–7 kg for chicken. Beef production is clearly more energy-intensive, less acidifying and has a lower eutrophication impact than producing broiler meat, but pork and beef have significant overlaps for all of these measures of environmental degradation.

A positive obverse of these burdens is that any reduction in meat production due to modified demand or a substantial improvement in the efficiency of meat output will result in lowering the extent and the intensity of all of these environmental impacts. But I will not open this section on environmental impacts of meat by looking at the effects of either extensive or intensive practices but by focusing on two essential indicators of global meat production, on the density of animal zoomass and on global totals of wild vertebrates and livestock. I will do this in order to demonstrate a little appreciated reality, the fact that domesticated animals now add up to the most massive category of heterotrophic macroorganisms in the biosphere.

Animal densities and aggregate zoomass

Few people realize to what extent domesticated animals dominate the vertebrate zoomass on the Earth. Zoomass is a term that could be applied, *sensu lato*, to all heterotrophic organisms (strictly speaking that would be, in size terms, from the smallest nanobacteria to the largest whales), but in this section I will use it, in *sensu stricto*, just for the mass of all terrestrial vertebrates except for the humans for whom anthropomass is a more suitable designation. High densities of domesticated animals and their dominance of terrestrial zoomass can be appreciated only when comparing them with the abundance and zoomass of wild animals living in natural, undisturbed, ecosystems.

Those densities correlate significantly with individual body mass. Damuth (1981) concluded that it is a simple log–log linear relationship with the exponent of –0.75, but Silva and Downing (1995) demonstrated that the exponent close to the expected –0.75 is valid only for the populations of intermediately sized (weighing between 0.1 and 100 kg) mammals, while for the heavier animals (100 kg to 3 t) the exponent was close to 0. This means that for every square kilometer, there would be only about ten animals weighing 1 kg and one with a body mass of 100 kg – but still one mammal weighing 1,000 kg. Not surprisingly, except for the largest animals, the actual densities show large (at least two orders of magnitude) density departures from this central tendency, but the total zoomass of individual ecosystems is always dominated by the largest terrestrial mammals.

As a result, the African elephant (*Loxodonta africana*), the largest surviving terrestrial animal, has peak East African densities in excess of 30 kg/ha (I use hectares because the densities of domesticated grazers are usually expressed per hectare rather than per square kilometer), while the densities of all small mammals (mostly rodents) rarely surpass 3 kg/ha and are usually much lower than 1 kg/ha (French et al. 1976; Smith and Urness 1984; Sinsin et al. 2002). The highest recorded zoomass densities were almost 200 kg/ha (20 t/km^2) in Uganda's Rwenzori National Park during the 1960s and just over 190 kg/ha in Tanzania's Manyara National Park (Coe et al. 1976). These were exceptional (and no longer prevailing) rates, while the densities for East Africa's grazer-rich savannas (including Serengeti, Masai Mara and Simanjiro Plains) ranged between 40 and 100 kg/ha (Schaller 1972; Kahuranga 1981; Ogutu and Dublin 2002). West African parks have lower herbivore densities of 10–20 kg/ha (Milligan et al. 1982; Sinsin et al. 2002).

Such wild animal densities were commonly equaled or surpassed by the herds of traditional pastoralists even in the more arid regions of East African savannas where the total zoomass of all domesticated herbivores (mostly cattle, also goats, sheep and/or camels) used to range between 10 and 15 kg/ha; in the more productive districts, it was above 50 kg/ha, and among the Maasai it was as high as 150 kg/ha (Dyson-Hudson 1980). Zoomass densities of modern beef cattle are considerably higher. In Brazil, cattle grazing on new grassland created by tropical deforestation has zoomass of 400–500 kg/ha, and similar densities apply to tall-grass prairie grazing in Oklahoma.

The EU's Common Agricultural Policy limits the stocking density to two livestock units, that is, around 1,000 kg/ha, and even higher rates can be reached seasonally when grazing on summer grasses or when grazing in fenced paddocks for just a few days before moving the animals to another site. In the former case, densities can be as high as 2,500 kg/ha (in South Africa's Natal); in the latter case, they can reach more than 15,000 kg/ha. And, as already explained, stocking densities in large feedlots, where tens of thousands of cattle spend months in confined feeding before slaughter, are commonly 20–30 m²/animal, that is, 200,000–300,000 kg/ha, and could be as high as 750,000 kg/ha. Obviously, such extraordinary densities are attainable only because of incessant, large-scale deliveries of high-energy and high-protein concentrate feeds.

Perhaps even more astonishing is the degree to which domesticated animals now dominate the terrestrial vertebrate zoomass. They belong to only seven genera of domesticated mammals – cattle (*Bos*), horse and donkey (*Equus*), water buffalo (*Bubalus*), pig (*Sus*), sheep (*Ovis*), goat (*Capra*) and camel (*Camelus*) – and four genera of birds – chicken (*Gallus*), goose (*Anser*), duck (*Anas*) and turkey (*Meleagris*). The Netherlands Environmental Assessment Agency reconstructed the livestock counts going back to 1890, and in 1900, there were roughly 450 million heads of cattle, 100 million horses, 195 million pigs, 150 million goats and 600 million sheep (HYDE 2012). A century later, the global cattle count, according to the FAO, was more than 1.65 billion, horses were down to 55 million, pigs (after nearly quintupling) were up to 905 million, and goat and sheep totals reached, respectively, 720 million and 1,050 million.

Using these totals and appropriate average body weights, I calculated that this domesticated zoomass rose from at least 170 Mt of live weight in 1900 – when it was more than four times larger than that of all wild terrestrial mammals – to at least 600 Mt of live weight in the year 2000 when it was approximately 25 times the mass of all wild land mammals (Smil 2011). And in 2000, cattle zoomass alone was at least 300 times greater than the zoomass of all African elephants whose zoomass was less than 2% of the live weight of the continent's nearly 300 million heads of cattle. And in the year 2000, despite humanity's unprecedented 20th-century increase from 1.6 to 6.1 billion people, live weight (in dry terms) of global anthropomass added up to no more than 110 Mt while cattle zoomass had surpassed 160 Mt, and dry weight of all domesticated animals reached nearly 250 Mt.

Given these realities, an extraterrestrial expedition to the Earth might logically report back that the planet's dominant life-forms (assessed strictly in terms of aggregate biomass) are mammalian quadrupeds of four species (cattle, pigs, goat and sheep) found in enormous concentrations across the ice-free landscapes. Of course, a significant share of the ruminant animals are kept as milk, rather than meat, producers, but they, too, are eventually slaughtered and eaten. And in countries with high level of meat production, livestock zoomass surpasses, or at least equals, the anthropomass, even if its count excludes all dairy animals.

For example, in 2011 the Netherlands had about 1.25 million heads of cattle for fattening or grazing, more than 12 million pigs and 1.1 million sheep and goats whose live weight was more than 50% greater than the mass of the country's nearly 17 million people (and the inclusion of dairy cattle would raise this multiple to 2). And in parts of the country, the highest density of domesticated zoomass has even partially upended the normal trophic pyramid as livestock weight is an order of magnitude greater than the combined biomass of all soil invertebrates. And even in the US, the world's most populous affluent country with more than 300 million people, meat-producing animals now weigh as much as the country's population, in average per capita terms the world's heaviest of all major nations due to an exceptionally high incidence of obesity.

Changing animal landscapes

Land impact of meat production is summarized by the fact that roughly a quarter of the ice-free continental surfaces are used for livestock grazing and that a third of all arable land is planted to feed crops. And grazing and feeding are also responsible for most of the tropical deforestation (forests turned into pastures and soybean fields) and for degradation of soils due to overgrazing and improper agronomic practices. While the new pastures created by tropical deforestation continue an old practice of extensive meat production (stocking densities are typically no more than 1 cattle head/ha), feed crops produced on arable land are an integral part of high-intensity agriculture, and their rising yields made it possible to reduce land devoted to feed production: America's livestock practices are land-sparing because average corn yield had more than quadrupled during the 20th century and hay productivity had doubled, from 2.7 to 5.5 t/ha.

Mass production of meat based on pasture, an option that has been best exemplified by practices on both American continents in general and by the Western US, Argentinian and Brazilian livestock operations in particular, is predicated on maintaining large herds of animals, and – given the inherently limited number of animal units that can be supported per unit of grassland without adverse impacts – this translates into extensive land claims and makes pastures the landscapes that are most obviously shaped by animals.

The global area of grazing lands grew slowly during the first 1500 years of the common era, roughly doubling to 220 Mha in the early 16th century (IIYDE 2012). By 1800, it surpassed 500 Mha, and a century later it reached nearly 1.3 Gha. About 250 Mha of this gain took place because of the westward expansion of the US, Canada and Australia, and another 200 Mha of grazing land were added in sub-Saharan Africa. An even slightly faster expansion rate continued during the 20th century when most new grazing land was opened up in Latin America where it had quadrupled since 1900 to about 560 Mha, mainly due to Brazil's post-1960 expansion of pastures that has no parallel worldwide. Pastures in sub-Saharan Africa had more than doubled during the 20th century to about 900 Mha. In the year 2000, permanent meadows and pastures occupied about 34 million km² (3.4 Gha) or about 26% of ice-free continental surface, or nearly as much as the world's remaining forests. In continental terms, Australia has the highest share of land in pastures (about 50%), followed by South America and Africa.

In contrast, about 11% of the Earth's ice-free surface (1.53 Gha) is under annual or perennial crops, with the continental shares ranging from more than 30% in Europe to less than 10% in Africa. This means that pastures account for more than half of all land used for food production on all continents with the exception of Europe, with shares as high as 80% in Africa and nearly 90% in Australia (FAO 2012). Traditionally, livestock grazing was an environmental adaptation making the best of limited resources. Even today, a properly practiced pastoralism should not be seen as an archaic way of food production with marginal economic returns and with a high potential leading to land degradation but as a rational, and potentially rewarding, option to exploit limited primary productivity of arid ecosystems (Rodriguez 2008).

But in general, pastoralism has been in retreat for centuries, and this process has accelerated during the past two generations due to the combination of population growth, overgrazing, agricultural encroachment and migration to cities. Livestock grazing is now overwhelmingly driven by the demand for meat, and its future extension will be ever more concentrated in only a few regions. That is why Asner et al. (2004) concluded that in most of the world's regions, the grassland expansion has run its course and that further forest conversions could create more grazing land only in parts of South America, Southeast Asia and Central Africa.

Countries with the largest areas of pasture are China (400 Mha), the US (nearly 240 Mha), Brazil (almost 200 Mha), Kazakhstan (185 Mha), Sudan (close to 120 Mha) and Mongolia (115 Mha). While both the grazing land and the total number of ruminants on pasture have been increasing, the numbers of true pastoralists have been declining (many should be now classified as agropastoralists), and even in the poorest African countries, pastoralism now accounts for less than 10% of the total GDP and for only a fraction of 1% in Australia, the country where the practice is relatively more important than in any other affluent nation. Extensive pastures with low plant productivity that cannot be raised without unrealistically expensive measures account for most of the existing grazing land, while intensive, managed pastures (many improved through drainage or irrigation, some receiving fertilizers) are concentrated in Europe, North America and some parts of Asia.

Perhaps the most spectacular sign of extensive pastoralism is a periodic burning of grasslands in many parts of the world. This long-standing practice is done in order to prevent conversion, the (re)establishment or encroachment of woody plants. The activity is marked by weeks and months of smoke and haze near the ground, and its extensive impact is now easily monitored by Earth observation satellites, perhaps most impressively as the annual fire season sweeps across much of sub-Saharan Africa. Other regions of frequent pasture burning are in parts of South America, Central and Southeast Asia, and Northeastern Australia.

Lauk and Erb (2009) estimated that fires burn annually 3.5–3.9 Gt of dry matter and that sub-Saharan Africa's grassland fires, at 2.2 Gt/year, accounted for as much as 60% of it. But this attribution is highly uncertain because of a great range of burning intervals: the most common recurrence is about 4 years; some pastures are left unburned for up to 20 years, others are burned annually. Interannual fluctuations of burned phytomass are thus considerable, and estimates of actual rates have differed by nearly an order of magnitude, ranging from just 0.22 to 1.85 Gt/year (Barbosa et al. 1999). The latest areal estimate is for the years 2001–2005 when 195 Mha of African grasslands were burned annually, releasing about 725 Mt C (Lehsten et al. 2009).

But these carbon emissions should not be added to the total of greenhouse gas emission generated by livestock because the carbon released by burning pastures will be promptly incorporated into a new season's grass growth; moreover, many pastures subject to natural fires set off by lightning and productivity of such fire-adapted ecosystems as tropical and subtropical grasslands might actually increase thanks to regular burning. As a result, it would be very difficult to quantify the negative environmental impact of periodic pasture burning. And it is no less difficult to quantify the impact of overgrazing of vulnerable grasslands, a long-standing problem that has only intensified since a rudimentary veterinary care improved the survival rates of pastured animals during the 20th century. At the beginning of the 21st century, this concern is greatest in many arid and semi-arid African landscapes and in their counterparts in Central Asia.

Pasture degradation can arise from soil compaction, loss of soil organic matter and nitrogen content, wind and water erosion, salinization and even from air pollution. Its consequences span a wide continuum of impacts ranging from those that can be managed (often as simply as by enforcing animal rotation among paddocks) to damage that takes out the affected land from production (after a severe water erosion turns a sloping pasture into a network of gullies). Obviously, unequivocal quantification of these effects has been elusive. Oldeman's (1994) often-cited global and continental totals showed that more than 20% of permanent pastures were degraded in the early 1990s, with most of the areas in Africa (2.4 Mkm²) and Asia (2 Mkm²). Estimates of land degradation caused by desertification have been particularly contentious, with some shares of the affected pastures being as high as 70% of the total area in arid regions (Dregne and Chou 1994). As a result, Steinfeld et al. (2006) offered a highly uncertain conclusion that anywhere between 20% and 70% of grazing land should be considered as degraded.

Quantification becomes easier when looking at the two most important landscape changes attributable to livestock raised for meat, to a large-scale conversion of forests (and, to a much lesser extent, also of wetlands) to create new pastures and new fields devoted to continuous cultivation of monocultures that now dominate feed supply. I will address the first concern in this section and deal with the second one separately later. Deforestation aimed at creating large areas of new pastures for extensive production of beef has been of the greatest concern in Latin America, and particularly in Brazil, as the process has been encroaching on parts of Amazonian rain forests, a unique set of ecosystems that is showing some signs of a transition to a disturbance-dominated system (Davidson et al. 2012).

Drivers of Amazonian deforestation have included road paving (both legal and illegal roads have opened the way to forest clearing) and logging, and since the early 2000s conversion to cropland (mostly for soybean export) has been an important component – but extensive, low-productivity cattle pasture has remained the single largest use of converted land, and although Brazilian deforestation rates may be off their historic highs, it is far from certain how much forest will be lost due to future conversions to pastures or to croplands.

Where livestock grazes alongside wild herbivores (as it does commonly in parts of Africa), the outcome of their competition will depend on the specifics of the contest, above all on the stocking densities of the principal species, on precipitation and on the frequency of fires set by pastoralists to prevent encroachment of woody phytomass. On African savannas, wild grazers can coexist with cattle thanks to resource partitioning as the animals choose different kinds, sward heights and ages of available grasses and other herbaceous phytomass. Cattle are normally considered to be non-selective grazers, but they may share grasses with wildebeest and zebras, with only seasonal overlaps: with zebras during the early wet season and with wildebeest during the early dry season (Voeten and Prins 1999).

Suttie et al. (2005) found that while wild grazers in Kenya's arid habitats avoid land that is heavily grazed by livestock, they mix with domesticated animals in semi-arid rangelands and may even graze close to settlements in order to get protection from predators. And evidence from the East African savannas shows that cattle compete with some wild herbivores during the dry season, but during the wet season grazing by wildlife actually improves the quality of forage for cattle (Odadi et al. 2011). Not surprisingly, heavy grazing exacts various costs, ranging from compaction of soils to nitrogen depletion of soils and grassland degradation; Cease et al. (2012) found that in north Asian grasslands, the latter process promotes outbreaks of *Oedaleus asiaticus*, a dominant locust in that environment.

Meat production in confinement has relatively small direct land claims as the areas occupied by pigsties, poultry barns and even enclosures for free-ranging chicken and large cattle feedlots add up to only a small fraction of land needed to produce concentrate feeds and forages and to graze the ruminant animals. A comparison makes this disparity clear. After weaning, a growing pig raised in confinement will be allotted less than 1 m² to reach its slaughter weight in less than six months to be followed by another animal. But that pig will consume about 300 kg of concentrate feed whose production would require – assuming a simple ration of 80% corn and 20% soybeans produced from US crops yielding, respectively, 10 and 2.5 t/ha – about 500 m² of cropland to produce. Consequently, even if we were to quintuple the space directly needed to raise a pig (in order to account for barn corridors, waste disposal, feed storage and yards), there would still be a difference of two orders of magnitude between direct space claims and the areas required to grow the feed.

Claims made by specific animal diets on arable land depend on a highly variable composition of feeds (determined by their shares of feed crops and food processing waste and use of other concentrates, and in the case of ruminants by the shares of forages, crop residues and other roughages) and on the origin of these feedstuffs produced with varying and fluctuating yields. The latter reality is best illustrated with corn, the world's leading concentrate feed: in 2010, its yields ranged from just over 4 t/ha for Southern Africa to 5.2 t/ha (global mean) to 9.6 t/ha in the US, and even in the US, annual averages fluctuate by as much as 10%. Given these realities, it is not surprising that land required to produce feed for specific meats in different countries at different times can range two- or even threefold.

Intensive production of feedstuffs

Cultivation of feedstuffs that are needed to raise the animals after weaning or hatching to their slaughter weight as fast as possible thus adds up to the single largest, and truly global, environmental impact of meat production. This is due not only to very large areas planted to feed crops but also to energy needed for their cultivation and harvests and to direct and indirect impacts on soils, water and the atmosphere. Given the intensity of feeding in modern mixed farming systems and, even more so, in confined large-scale operations, and given the high rates of per capita meat consumption in affluent countries, it is not surprising that the cultivation of feed crops aims for high yields and is done by intensive methods, that feed rather than food crops now occupy most of the cultivated land in countries with plentiful farmland, that even globally those crops now account for a third of all cultivated land (i.e., for almost 500 Mha) and that the countries with limited farmland must resort to substantial imports of feedstuffs.

By far the best example of the latter necessity is Japan, now the world's largest importer of feedstuffs. American feed corn became the country's most important imported grain already during the late 1960s, and in the year 2000 Japan imported more than 6% of America's corn crop whose cultivation required about 1.8 Mha, an equivalent of nearly 40% of Japan's arable land or roughly all of the land planted to corn in South Dakota (Smil and Kobayashi 2012). In addition to corn, Japan also imports soybeans (both for food and feed) and roughages. South Korea is now the world's second largest affluent importer of feed grains, while China has become their leading low-income importer.

The list of feedstuffs used in modern meat production is a long one, including rich sources of carbohydrate (corn, sorghum, wheat, barley, potatoes, cassava) and protein (legume seeds but most commonly various oil cakes) as well as many kinds of milling and processing by-products (cereal brans, vegetable- and fruit-processing wastes) and, for the ruminants, forages ranging from leguminous cover crops (alfalfa, clovers, vetches) to various silages and hays. But the imperatives of optimized feeding (combination of carbohydrates and protein now usually supplemented by micronutrients) at minimized expense have narrowed the most common choice of feedstuffs to a small group of species of high-yielding (in absolute or relative sense) crops whose combination now dominates commercially formulated or custom-made feed mixtures.

Two crops are now undisputed leaders, corn as by far the most important source of carbohydrates and soybeans as the leading supplier of protein. Their importance led to an early development of transgenic varieties, and now most of the world's commercial feed harvest of corn and soybeans comes from genetically modified plants. In order to maintain continuous high yields of their monocultures or limited rotations, these crops require large inputs of fertilizer, particularly nitrogen (even soybeans now often receive supplementary nitrogen applications), and have become a major reason for substantial leaching of nitrates as well as for losses of organic nitrogen in eroded soil. This makes them leading contributors to nitrate water pollution, stream, pond and lake eutrophication and to the evolution and persistence of some extensive dead zones in coastal waters.

Corn, the principal ancient grain of the Americas, has been America's leading crop throughout the 20th century, but its importance rose after the introduction of high-yielding hybrid varieties during the 1930s as the yields increased from less than 2 t/ha in 1940 to 3 t/ha in 1960 and about 8 t/ha by the year 2000 (Kucharik and Ramankutty 2005). Since 2004, America's corn yields have averaged more than 9 t/ha, and in the Corn Belt states they have been above 10 t/ha. The country exports about 20% of the crop, and, as already noted, Japan is the largest importer. Most of the US corn is grown in two-year rotation with soybeans or in two- to three-year rotations with alfalfa, cotton and other crops.

Rotation with soybeans is not as beneficial as might be assumed because the high yield (i.e., high grain/straw ratio) of improved soybean varieties means that this nitrogen-fixing plant may be a net user of soil nitrogen rather than a source of its replenishment. Moreover, soybean is also a row crop, and before its canopies close the exposed soil is vulnerable to high rates of rain erosion in sloping fields. In contrast, alfalfa leaves behind a large surplus of fixed nitrogen for the subsequent crop, and it provides excellent anti-erosion protection. Unfortunately, its cultivation has been declining or stagnating.

There are three reasons why my quantification of energy costs of leading feed crops will focus on the US: the country has the world's highest ratios of feed/food field crops; it is a major exporter of feed corn and soybeans (as a result, it suffers the associated environmental impacts that are avoided by importers of US feedstuffs); and its agricultural statistics are excellent. Energy inputs into the production of field crops consist of two major components, of direct energy expenses for fueling tractors and other machinery (particularly for pumps used for irrigation and sometimes for drying the harvested crop) and of indirect energy costs embodied in farm machinery, applied fertilizers, pesticides and herbicides. All of these costs have their own ranges, and when combined with different agronomic practices (both in time and space) and with specific crop needs, they will result in rates that may differ substantially from those for other crops or for the same crop grown under different conditions.

This rather wide range of possible outcome must be kept in mind because the energy cost of feed is almost always the single largest component of the final energy cost of edible product. Energy analyses of North American and European crop production found rates between 8 and 15 GJ/ha for dryland cereals as well as for soybeans and at least 20 GJ/ha for high-yielding Corn Belt corn grown without irrigation and more than 40 GJ/ha with supplementary watering (Smil 2008). When expressed as energy costs per unit of feed, these values prorate to as little as 1.5 GJ/t and to as much as 4 GJ/t. Ten different studies of US corn, the leading concentrate feed, put its energy cost between 1.5 GJ/t and just over 3 GJ/t (Kim and Dale 2004; Landis et al. 2007). Breakdown of these costs indicates where the greatest savings could be made with future rationalized production.

Modern corn production is completely mechanized, and tractors powered by diesel engines will need 600–1200 MJ/ha of moldboard plowing, about 250–300 MJ/ha for soil finishing, only 100–200 MJ/ha each for planting and cultivating, 150–350 MJ/ha for fertilizer application and 250–500 MJ/ha for grain harvesting. All field tasks from seeding to harvesting will thus claim at least 1.5–2.5 GJ/ha. Hay cutting requires 100–150 MJ/ha, alfalfa baling needs about 50 MJ/t of crop, and the overall energy cost of forage harvesting is no less than 500 MJ/ha (Gelfand et al. 2010). Gasoline-powered tractors would need roughly one-third more, while no-till cultivation cuts down fuel needs by 65–85%. With production costs of 70–120 GJ/t of tractors and major implements, the annual cost of energy embodied in machinery (typically prorated over 10–20 years of expected service) is broadly comparable, ranging (with maintenance and repair costs included) between 1 and 5 GJ/ha.

These costs are dwarfed by the need to irrigate (whenever crop's evapotranspiration and leaching is greater than the sum of stored soil water volume and precipitation): their five key determinants are the volume of water to be delivered; the mode of delivery (surface or wells); and engine, pump and irrigation efficiencies (the first typically between 25% and 35%, the second between 50% and 75%, the last as low as 30% with traditional furrows and easily 75% with properly operated center-pivot sprinklers). Permutations of these conditions will yield rates ranging from less than 1 GJ/ha to more than 10 GJ/ha, the latter value surpassing all other energy needs in crop production.

High-yielding corn requires large fertilizer inputs: US national averages, heavily influenced by high Corn Belt yields of around 10 t/ha, prorate (reduced from actually applied compounds to elemental rates) to about 150 kg of nitrogen, 30 kg of phosphorus and 80 kg of potassium/ha (USDA 2012c). Phosphorus and potassium are not energy-intensive inputs (their mining and processing requires, respectively, around 20 and 10 GJ/t of nutrient), but nitrogenous fertilizers based on Haber–Bosch synthesis of ammonia require natural gas both as feedstock and the fuel for the synthesis; with subsequent conversion to urea and with packaging and distribution, this comes to at least 55 MJ/kg. Fertilizer applications required to produce American feed corn thus have embodied energy content of about 10 GJ/ha.

Both herbicide and insecticide applications began in the US during the mid-1940s and elsewhere during the 1950s. Synthesis of their active ingredients is fairly energy-intensive, requiring 100–200 MJ/kg, and formulating, packaging and marketing pushes the cost to 200–300 MJ/kg (Unger 1997). But, unlike fertilizers, these compounds are applied in small quantities (just 1–2 kg/ha), and hence their relative energy cost is less than 1 GJ/ha. Producing American feed corn thus takes at least 15 GJ and up to 20 GJ of direct and embodied energy inputs per hectare, and with good yields around 10 t/ha, this prorates to 1.5–2 GJ/t. Obviously, no-till cropping and rotation with alfalfa will lower these rates.

Soybeans, traditionally cultivated as a protein-rich food crop in East Asia, have become by far the most important source of feed protein, with some cultivars containing more than 40% protein compared to just 10% in corn and 15% in wheat. Because of their association with nitrogen-fixing *Rhizobium* bacteria, soybeans need no or little supplementary N fertilization, and in recent years only about 20% of the US crop received less than 30 kg/ha; but average P and K applications are similar to that for corn, averaging, respectively, 25 and 75 kg/ha (USDA 2012c). The total weighted energy cost of fertilizers is only about 1.5 GJ/ha, and the US soybean production consumes typically no more than 6 GJ/ha.

With recent yields fluctuating between 2.5 and 2.9 t/ha, that prorates roughly to 2–2.5 GJ/t, or 25–35% more than for feed corn (but soybeans have at least three times as much protein). But a recent Brazilian study put soybean production costs in two principal growing regions, Center West and South, much higher at, respectively, 4.8 and 3.9 GJ/t, while several earlier evaluations ranged between 1.2 and 3.8 GJ/t (da Silva et al. 2010). Transportation costs of Brazilian soybeans may be as high as, or even higher than, the production cost: 3 GJ/t for road transport and 1.5 GJ/t for shipping to Europe (Rotterdam). The total of 4.5 GJ/t is only slightly lower than the cultivation demand of 4.8 GJ/t.

Harvested forages, essential for proper nutrition of ruminants, include single-species or mixed hay and leguminous cover crops, above all alfalfa, clover and vetch. Aggregate figures for hay harvests produced on arable land are available only for a few countries: their recent rates were about 140 Mt (fresh weight)/year in the US and close to 20 Mt in France, with the global output on the order of 1 Gt/year. Hay has about 10% moisture, while corn silage contains 75% water.

Energy cost of roughages differs widely. Obviously, improved pastures (this may include drainage and periodic fertilization) require more energy than natural grasslands (where the main external energy cost is fuel for harvesting). American hay needs 2–3 GJ/t; similarly, a study of bromegrass in Iowa (adding up energy costs of seed, nitrogen, diesel fuel and fencing required to establish a pasture and energy cost of annual maintenance) prorated on an annual basis to 3 GJ/ha. Energy cost for alfalfa, by far the most important US fodder crop, is fairly similar for establishing new fields (on the order of 10 GJ/ha) and for harvesting and delivery to feedlots (mowing, baling and transporting bales). Alfalfa obviously needs no nitrogen, but it often receives potassium (about 100–120 kg/ha) as well as supplementary liming amounting up to 200 kg/ha (Kim and Dale 2004).

As with grain crops, the main difference in the overall energy cost is primarily due to rainfed and irrigated cultivation and secondarily to irrigation using surface water or water pumped from wells. Annual rainfed production of a single cutting may cost only between 10 and 12 GJ/ha or just 1–1.5 GJ/t, while irrigated cultivation would claim easily 2.5 and as much as 4 GJ/t (Heichel and Martin 1980; Tsatsarelis and Koundouras 1994). I must reiterate that specific analyses may result in much higher (particularly for crop irrigated with water drawn from deep wells) or somewhat lower energy costs.

Given the relatively large ranges of energy subsidies for the three major categories of cultivated feed, I will use an overall average of 3 GJ/t (3 MJ/kg) of feed for the first-order approximation of livestock's overall energy subsidies. In 2010, animals consumed about 800 Mt of grain feed, with nearly 40% going for the production of milk, eggs and fish; roughly

500 Mt of concentrates (dominated by corn) used to produce meat thus required on the order of 1.5 EJ of external energy subsidies (mostly for field machinery and irrigation and for synthesizing fertilizers). For comparison, 2010 global primary consumption of commercial energy was about 500 EJ, which means that the cultivation of feed grains was only about 0.3% of the total. This conservative value underestimates the total global energy needed to produce animal feedstuffs – but it is of the right order of magnitude.

As already explained, the most intensive cattle feeding requires about 8 kg of dry matter feed/kg of gain. With 3 MJ/kg, the external energy subsidies could be as low as 25 MJ/kg of gain, but when the feed comes from irrigated land, the costs may be 10 MJ/kg of feed and the subsidies may be on the order of 100 MJ/kg of beef. Several American studies that included the energy cost of breeding animals indicated a range of 60–150 MJ/kg of feedlot beef. Energy needs for housing the animals, processing their feed, delivering water and managing manure are much more variable. For traditional grazing they may be negligible, but modern North American practices are fairly energy-intensive because of fencing, periodic pasture improvement (irrigation, drainage, fertilization) and liquid fuels needed for pick-up trucks and hay harvesting; as a result, weaner calves may need anywhere between 18 and 130 MJ/kg of gain. Energy for feedlot management (typically 3–10 MJ/kg of gain) is a small fraction of energy required to produce feed, but electricity needs for air-conditioning of poultry CAFOs in warm climates may be considerable.

In contrast, data available for energy cost of meat processing (slaughtering, evisceration, cutting, deboning and water cleaning, cooling, freezing) have a fairly restricted range, with energy costs mostly between 1.5–3 MJ/kg of chilled meat and 3–6 MJ/kg of frozen meat (Ramírez 2005). But, once again, transportation costs vary substantially, depending on the mode (truck, train, ship), distance and meat temperature (chilled or frozen), with the common range being between 2 and 10 MJ/tkm (or 2 and 10 kJ/kgkm). The only valid generalization is that the energy cost of feed needed to produce red meat dominates the overall cost of production–slaughtering–processing–delivery chain.

Water use and water pollution

Livestock's impacts on water resources and water quality are obvious. Direct water demand comes largely from three different needs: all domesticated animals must have regular supply of water for drinking; while excreta of grazing animals will recycle water and the nutrients to benefit grassland soils, removal and disposal of wastes of animals raised in landless, concentrated feeding operations need water for washing and cleaning and also for mixing feeds; and relatively large volumes of water are required by slaughterhouses and meat-packing enterprises. But in all instances, this direct water demand is dwarfed by indirect needs (virtual water) to grow animal feedstuffs, be they forages or concentrate feeds.

Traditional concerns about water pollution consequences of meat production focused on animal manures: on their removal, on contamination of nearby sources of drinking water, on leaching and overflow from liquid waste ponds, and on nitrogen losses from field applications of those organic wastes, with excessive nitrate concentrations in water seen as a major risk to human health (WHO 2007). These concerns remain, but given the intensive cultivation of feed crops, a comprehensive analysis of water pollution attributable to meat production must also consider the losses of fertilizer applied to feed crops whose most worrisome impact is not excessive contamination of drinking water but rather eutrophication of sensitive ecosystems, especially coastal waters that may be many hundreds of kilometers from the source of leached and eroded nitrates. An argument can be made that these effects are both more important, and certainly more intractable, than the direct and indirect water footprint of meat production.

Quantifying livestock's water needs cannot be done with great accuracy even for drinking water, a relatively small direct demand whose calculation is based on well-known individual requirements. Uncertainties arise when multiplying specific animal numbers by average daily water requirements whose rate obviously varies with air temperature and overall exposure. Naturally, lactating animals have considerably higher need than adults of the same age raised for meat. Typical rates for air temperatures no higher than 25°C are between 35 and 40 L/day for a head of beef cattle, 7 and 8 L/day for a growing pig, and adult broilers need 20–35 L/day for every 100 heads (Schlink et al. 2010). Steinfeld et al. (2006) put the annual drinking water requirements of the world's livestock at 16.26 km^3, with cattle and water buffaloes taking nearly 80% and with poultry's aggregate surpassing the demand by pigs; on the regional basis, predictably, South America and South Asia dominate. A more recent, and more detailed, assessment by Mekonnen and Hoekstra (2010) put the total nearly 70% higher, at 27 km^3.

Service water needs (washing and cleaning of animal and barns) range from nothing for suckling animals to about 5 L for grazing cattle, around 10 L for beef cattle in CAFOs and 50 L for mature pigs raised in confinement; for growing broilers, the average is about 9 L/100 birds

(Chapagain and Hoekstra 2004). Using these rates, Steinfeld et al. (2006) calculated the annual global service water needs to be less than $7\,km^3$ with pig raising being the largest (nearly two-thirds of the total) consumer – but Mekonnen and Hoekstra (2010) ended up with nearly three times as much, $18\,km^3$. These substantial differences in direct (drinking and service) water use (amounting to between 25 and $45\,km^3/year$) are irrelevant to any overall summaries of livestock's water consumption as the estimates of water evapotranspired during the growth of feed crops are two orders of magnitude higher.

This indirect consumption is now commonly called virtual water; its concept was introduced during the 1990s (Allan 1993) following a previously used distinction between actual energy content and embodied (virtual) energy cost of commodities that can range from less than $5\,MJ/kg$ of bricks to more than $15\,GJ/kg$ of microchips. Because photosynthesis involves such an extremely lopsided trade-off between CO_2 and H_2O, even corn, a C_4 plant following a more efficient photosynthetic pathway, requires 400–500 mol H_2O/mol of CO_2 fixed, and the less water-efficient plants following C_3 metabolic pathway (including such major feedstuffs as barley, wheat and leguminous grains) need 900–1,200 mol (and even up to 4,000 mol) of H_2O to fix 1 mol of CO_2 (Smil 2008).

In reality, this means that production of grain corn for feeding claims 1,000 L of virtual water/kg of grain, and that the rates for barley, wheat, oats and soybean cakes are close to or above $1,500\,L/kg$ (for comparison, virtual water content of steel can be as low as $6\,L/kg$). Only a very small fraction of all evapotranspired water is actually incorporated in harvested parts: both cereal and leguminous feed grains are harvested when their moisture content drops below 20–25% and, in order to prevent excessive spoilages, are marketed at between 10.5% and 14.5% of moisture. For example, 1 t of feed corn whose growth evapotranspired 1,000 t (1,000 m^3) of water during its 150 days of growth contains only about 140 kg of water.

Approximate regional or national water requirements of feed crops can be calculated by inserting specific data into models of crop photosynthesis: evapotranspiration rates can range from less than 0.5 mm/day during winter months in temperate regions to more than 6 mm/day in arid subtropical countries, and actual volumes of water processed by identical crops grown in the same region can differ by as much as a factor of 2 depending on cultivars, soil qualities and agronomic practices. Low feed conversion efficiencies of meat-producing animals multiply these virtual water needs, and hence even the most efficient of all mass meat-producing systems, that of raising American broilers fed optimal diet, require more than 2,000 L of evapotranspired water/kg of meat.

Other meat rates calculated by Hoekstra and Chapagain (2007) are substantially higher: they put the world average for pork at about 5,000 L/kg, and their mean for beef was approximately 15,000 L/kg of boneless beef. Not surprisingly, detailed studies of specific settings and practices reveal rates that are very different from these often-cited averages, and this variability must be kept in mind in order to avoid commonly repeated assertions about meat, and particularly beef, production as a major driver of water scarcity. Comparison by Ridoutt et al. (2012) showed previously published water use values for bovine and ovine meat ranging from less than 4,000 L/kg to as much as 200,000 L/kg, with the most frequent values between 9,000 and 16,000 L/kg – but Beckett and Oltjen (1993) concluded that the typical cost of boneless beef production in the US is much lower than previously suggested, with the average requirement of only 3,683 L/kg.

Other revealing comparisons of meat's water intensity with the footprints of other common foodstuffs are done on energy or protein basis. Averages calculated by Mekonnen and Hoekstra (2010) are about 2.4 L/kJ of beef, 0.5 L/kJ of pork and 0.7 L/kJ of chicken; despite a higher feed conversion efficiency, chicken tops pork because of a higher share of grain concentrates in typical feed rations; in terms of protein, beef with 112 L/g is twice as water-intensive as pork with 57 L/g and more than three times as water-intensive as chicken with 34 L/g. Obviously, food energy from cereals and tubers is much less water-intensive (mostly just around 0.1 L/kJ), and water intensity of plant proteins is also significantly lower (no more than 20 L/g for cereal and legume proteins), but, of course, the quality of plant proteins is inferior to complete proteins in meat.

The most disaggregated calculations of the global water footprint of animal production, made by Mekonnen and Hoekstra (2010), assume surprisingly high averages of feed conversion efficiencies – for example, for beef the authors use (all in kg of dry feed mass/kg of output) 70.2 for grazing, 51.8 for mixed farming and 19.2 for landless operations, resulting in an overall average of 46.9 – and use weighted means of domestically produced and imported feeds for national feed components. Their grand total of water evapotranspired in animal feed production (annual average for the period of 1996–2005) came to 1,463 km³, and the aggregate for the four major feed crops (corn, soybeans, wheat and barley) was about 874 km³. For comparison, Steinfeld et al. (2006) put the evapotranspiration of the same four feed crops nearly 30% higher at 1,103–1,150 km³.

Mekonnen and Hoekstra (2010) also calculated that an additional 913 km³ were evapotranspired during the growth of pastures, and with direct (drinking and service) water demand at just 46 km³, their total water footprint of animal production reached 2,422 km³, or 1,931 km³ after subtracting the claims by dairy cattle and layer chicken. Consequently, global meat production has been claiming about 23% of all water used in agricultural production, about 8,300 km³, with annual crops accounting for nearly 90%. The study has also apportioned the overall demand among the three different sources of water, a critical separation that underscores the dominance of green water (drawn from soil and evapotranspired during the production of forages and feed crops) and relatively small requirements for blue (originating in surface storages or underground aquifers and used for drinking and services as well as for irrigation) and gray water (that becomes polluted during crop cultivation or livestock production, most often by nitrates and residues from pesticides and herbicides).

According to Mekonnen and Hoekstra (2010), about 87% of all water needed for livestock production is green water required to grow forages and concentrate feed crops, and the remainder is about equally divided between polluted gray water and withdrawals of blue water (each roughly 150 km³/year). This distinction is particularly important when looking at meat's water footprints at a national level. Even a relatively large water claim is of little concern in a country where all but a tiny fraction of it is green water supplied by abundant precipitation, while even a relatively small claim (in absolute terms) may be worrisome in a country where most of it is blue water drawn expensively from deep, and declining, aquifers to irrigate cultivation of feed crops (be it alfalfa or corn).

Making this distinction is essential in order to avoid condemning all meat as an excessively water-intensive commodity. As in so many similar instances of a system-wide life-cycle analysis, the setting of system boundaries and a uniform use of consistently defined variables make an enormous difference (Ridoutt and Pfister 2010). Green water is not directly accessible by humans, and until it turns into blue water, it does not contribute to flows that maintain the viability of freshwater ecosystems – and that is why Ridoutt et al. (2012) argued against including it in their analysis of water cost of pasture-raised lamb cuts produced in Victoria, Australia, and exported to California. With the exclusion of green water, overall consumption was just 44 L/kg of cuts (bone in), a result they think to be representative of other low-input, non-irrigated grazing operations.

Moreover, it is also necessary to remember that green water used in feed production is not consumed: once the evapotranspired water reenters the atmosphere, its subsequent fate can range from getting rapidly condensed on a nearby plant or soil to a long-distance (hundreds to thousands of kilometers downwind) transport before it gets precipitated on land or water. As a result, the same water molecules that were a part of producing Midwestern corn to feed pigs in Iowa may help to grow, just a few hours later, soybeans in Illinois that are eventually exported to Japan or, a week later, grass grazed by beef cattle in Wales.

And why should evapotranspiration resulting from cattle grazing on an African savanna be considered a burden (water cost) while evapotranspiration attributable to concurrent wildlife grazing should carry no negative connotation? And, obviously, evapotranspiration would take place even in grasslands devoid of any herbivores, and water "lost" that way has no real alternative use, being, one might say, the greenest of the green category. These realities provide a perfect example of the necessity to subject various indicators of environmental impact to critical deconstruction: often they do too little (most often by ignoring quality); sometimes they do (as in this case) too much.

Water pollution caused by livestock has many components, with deleterious effects ranging from very high oxygen demand for the decomposition of wastes to contamination with heavy metals, but in mass terms it is dominated by excessive releases of reactive nitrogen, above all of highly water-soluble nitrates originating in manure and in fertilizers used to grow feed crops (Smil 2001b). Point sources of this pollution range from individual leaking pig sties to the largest CAFOs and slaughterhouses; improved pastures and intensively fertilized fields used to grow feed crops are the largest non-point sources of reactive nitrogen.

Domestic animals are inefficient processors of feedstuffs in general and of nitrogen in particular. Even such relatively efficient protein convertors as fast growing young pigs excrete more than two-thirds of ingested nitrogen, but in relative terms (per kilogram of live weight) beef cattle are the largest producers of feces and urine (only about 5% of ingested nitrogen is retained in meat) and broilers are the least offensive (retaining up to about 40% of ingested nitrogen). As a result, only a small share of feed crop nitrogen ends up in consumed food: for example, according to Bleken and Bakken (1997), nitrogen retained in Norway's animal foods was just about 20% of the original input.

Three trends have worsened the nitrogen pollution from livestock: increasing separation of livestock production from cropland, ubiquitous applications of inexpensive synthetic nitrogenous fertilizers and the growing size of CAFOs have turned formerly valuable organic fertilizer into wastes that are difficult to dispose. Quantifications of animal waste production are not very reliable because the output rates vary according to breeds, body weights, feed qualities and individual health status. Dairy cows are by far the largest producers of urine and feces when measured in relative terms (excreta per unit of live weight), followed by beef cattle, chicken and pigs. Pits, tanks and lagoons are used in warmer climates for treating diluted manure (stabilizing organic matter, reducing the mass, venting ammonia) for a longer period of time before spraying on cropland. Nitrogen content of fresh wastes is low (just 0.5–1.5%), and their handling, transportation and application costs are high in comparison to much more concentrated synthetic compounds.

The annual manure production of a broiler is about 5 kg/year, for a pig in finishing operation it is nearly 550 kg/year and for a feedlot steer it is roughly 4,500 kg/year (MacDonald et al. 2009; Key et al. 2011). The last rate means that a feedlot of 50,000 animals will produce annually 225,000 t of manure that will contain about 1,250 t of nitrogen and 165 t of phosphorus, and it would need roughly 9,000 ha of corn field if all that manure would be recycled to fields. I estimated that during the mid-1990s, livestock excreted about 75 Mt N and that almost half of that total was recycled in manures (Smil 1999). The FAO's global estimate for 2004 was 135 Mt N, with cattle contributing 58%, pigs 12% and poultry 7%.

On traditional mixed farms, virtually all manure was recycled as it was usually the single most important source of nitrogen and phosphorus for crop cultivation; this practice persisted for decades after the introduction of synthetic fertilizers, and in all poorer regions of the world it is still the norm as manures complement synthetic fertilizers. But in the West in general, and in the US in particular, the practice began to decline with the adoption of landless livestock production and became only a fraction of the former extent with the widespread shift to CAFOs. But the practice remains desirable as organic nitrogen in manures leaks at lower rates than inorganic fertilizers and helps to create better soil structure.

Oomen et al. (1998) argued that, ideally, manure produced by the EU's pigs and poultry should be returned to croplands that had produced the feed, a solution that would also improve animal welfare by giving the animals more living space. On the other hand, manuring can lead to a build-up of excess nutrients, and it is a major source of volatilized ammonia. And recycling manures produced in CAFOs means higher energy use for distributing wastes from those large facilities over larger areas, making manures more expensive than synthetic fertilizers. Dispersing the animals into more mixed-farming operations could not be easily accomplished with so few people now working in agriculture.

As a result, in recent years only about 5% of all US cropland is now fertilized with manure with, not surprisingly, corn receiving about half of all those applications (MacDonald et al. 2009). But this trend may have run its course, as stricter environmental regulations (adopted in response to water pollution and high risks of spills from giant waste lagoons adjacent to CAFOs) will lead to more recycling. At the same time, intensive manuring in some densely populated regions led to legislative limits on this ancient practice: EU rules now limit the number of manure-producing animals per hectare of land available for manure spreading in order to prevent the saturation of croplands with organic wastes.

Besides excessive N and P loadings, intensive manuring can also lead to accumulation of heavy metals including Cd, Co, Cu, Fe, Se, Zn (manures are now usually their second largest input following atmospheric deposition) that were added to animal feed as micronutrients and growth promoters. Animals absorb typically 5–15% of their inputs and excrete the rest. Copper and zinc levels are usually highest, and cadmium levels are typically two orders of magnitude lower – but unlike the first two elements, cadmium is not an essential micronutrient; it is highly toxic and its accumulation in plant tissues is a clear health hazard. In addition, pesticides used to control insects in poultry houses and antibiotics used in all forms of animal husbandry can be found in manures, but we know little about the fate of these chemicals, or their residues, after manure applications. Wastes that are not recycled to fields could be used by CAFOs to generate methane through anaerobic fermentation generation, or they can be turned into feed ingredients.

But by far the greatest nitrogen losses causing water pollution and eutrophication are the result of inherently low rates with which the nutrient is absorbed from applied inorganic fertilizers. This interference in the global nitrogen cycle began to rise slowly during the 19th century with the mining of Chilean nitrates, later with the use of ammonia sulfate and cyanamide; by 1909, Fritz Haber demonstrated the synthesis of ammonia by combining air-derived nitrogen and hydrogen, but mass production of ammonia-based fertilizers took off only after WW II (Smil 2001b). Nitrogen fertilizers applied to feed crops and improved pasture now amount to about half of the nutrient used in the US and French farming, and close to 20% of all applications in China. Lower ammonia prices led to higher applications of inorganic nitrogen compounds (dominated by urea, nitrates and ammonia solutions) that are accompanied with significant losses of reactive nitrogen.

While some crops receiving moderate amounts of nitrogen can recover as much as 60–65% of the applied nutrient, typical recovery rates in temperate agroecosystems are below 50%, in the tropics and in arid regions even less than 30%. Nutrients' recovery rates for the Western grain crops grown for animal feed range mostly between 35% and 50% are lower for Asian and Latin American corn, but they may be as high as 75% for fertilized cropland forages. Lost nitrogen can follow several very different paths, and apportioning their typical shares in terms of large-scale averages is a matter of educated guesses. The most desirable loss is a complete denitrification, that is, a bacterially mediated return of reactive nitrogen into inert atmospheric N_2 whose rates vary widely. Volatilization of ammonia from plant tops and from surface-applied synthetic compounds and manure is also highly variable.

Soil erosion and leaching are the two pathways transferring nitrogen to waters, but reactive compounds from atmospheric deposition also end up eventually in streams, lakes and coastal waters. In the past, the principal concern about the leached nitrates from synthetic fertilizers and manures was due to their potential impact on human health. While nitrates are not toxic to humans in concentrations usually present even in polluted waters, their bacterial conversion to nitrites carries some health risks (Hoering and Chapman 2006). Even so, nitrates ingested by humans come mostly from vegetables rather than from water, above all from beets, celery, spinach, lettuce and radishes, and safe limits for drinking water have been widely adopted.

As a result, the major concern about reactive nitrogen in waters whose primary productivity is usually limited by shortages of nitrogen and phosphorus is now their eutrophication effect: excessive presence of the nutrient is manifested by algal growth and by reduced oxygen content of water due to the decomposition of dead phytomass, and this may lead to dead zones, that is, to die-off of bottom-dwelling organisms in shallow coastal waters (Orive et al. 2002). A large eutrophication-induced hypoxic zone in the Gulf of Mexico has received a great deal of attention, but it has numerous counterparts on four continents, and the problem (caused also by urban and industrial discharges) is still intensifying in many regions.

Xue and Landis (2010) quantified nitrogen flows during production, processing, packaging and distribution of red meat and chicken. Not surprisingly, they found that the red meat has by far the highest eutrophication potential, about 150 g of nitrogen equivalent/kg of meat compared to less than 50 g for a kilogram of chicken and less than 3 g to supply a kilogram of cereals. Nitrogen losses in cropping and meat and dairy production can be reduced by better management, but only in an incremental way, and the eventual practically achievable cuts will be on the order of 10–15% (Oenema et al. 2009).

Meat and the atmosphere

Offensive odors were the most common airborne consequence of animal husbandry in traditional settings as pungent ammonia emanated from barns, sties and coops, but with relatively small concentrations of animals, these impacts were spatially limited mainly to enclosed structures or to areas in the immediate vicinity or little further downwind. Inevitably, much larger concentrations of animals in modern landless operations have magnified these problems, and ammonia, other odorous compounds (including hydrogen sulfide and dimethyl sulfide) and bioaerosols (often laden with pathogens) have lowered air quality around CAFOs and have exposed many people to objectionable air pollutants (NRC 2002).

In global terms, ammonia emissions from livestock had more than tripled during the 20th century (Galloway et al. 2004). Emissions from areas with high density of cattle or pigs, and even more so with a multitude of large CAFOs, are substantial enough to create areas of significantly higher atmospheric NH_3 concentrations and create regions of elevated ammonium ion (NH_4^+) deposition for tens to hundreds of kilometers downwind. This process is perhaps best illustrated with the maps regularly published by the National Atmospheric Deposition Program that show the highest levels of ammonium ion wet deposition in the Corn Belt, centered on Iowa and reaching into southern Minnesota, eastern South Dakota and Nebraska, northern Missouri and western Ohio and Illinois (NADP 2012). The other important source of reactive nitrogen in the atmosphere that is attributable to livestock (and that also contributes to Corn Belt's emissions) is NH_3 volatilization from synthetic fertilizers (particularly from urea) used to grow feed crops.

While ammonia emissions have been a part of long-standing concerns about environmental impacts of meat production, livestock-related emissions of greenhouse gases are a fairly new concern whose origins date back to the late 1970s. The concern is due to a fact that both methane and nitrous oxide – the two gases whose global emissions are dominated, directly and indirectly, by livestock – are more potent absorbers of the outgoing long-wave radiation than carbon dioxide (by far the most abundant anthropogenic greenhouse gas), and hence, even in small concentrations, they are relatively large contributors to the global warming effect.

And these concerns became even more prominent when some calculations showed a surprisingly large contribution of livestock to the global generation of these compounds whose rising tropospheric concentrations have already helped to increase the Earth's average temperature during the past 150 years and whose further increases pose a risk of planetary warming whose degree would be unprecedented in the history of our species (IPCC 2006; Steinfeld et al. 2006; Goodland and Anhang 2009). I will review these findings and explain and criticize some of their conclusions, mainly by stressing the facts that the accounting for livestock-related emissions of the three principal greenhouse gases (CO_2, CH_4, N_2O) involves a combination of some fairly reliable quantifications and of many more uncertain estimates whose ultimate aggregates are also critically dependent on the boundaries chosen for those analyses.

Domesticated animals are responsible, directly and indirectly, for significant releases of all three major greenhouses gases, and as each of them is generated by more than a single and readily quantifiable process, their estimates have to consider an array of sources. But before that is done, they have to set some arbitrary analytical boundaries on what will get included and then they have to resort to many concatenated assumptions in order to calculate national, continental and global aggregates of specific emissions. As in the case of meat production's impacts on the hydrosphere, the following global (as well as regional and national) estimates and quantifications include the contributions by dairy cattle.

Given the inherently uncertain nature of these kinds of quantifications, satisfactory corrections for animals raised only for their meat could be done by simply reducing the aggregate numbers by at least 20% and by as much as 30%. In fact, the inclusion or exclusion of dairy animals may make only a marginal difference as uncertainties on the order of 20–30% are the minimum errors to be expected when making the global estimates of greenhouses gas emissions attributable to livestock. Preparing these estimates is complicated not only due to large variability of many specific fluxes but also due to the necessity of including three different gases that originate from a number of disparate sources and by the imposition of arguable boundaries on the analyzed emission sets.

Enteric fermentation of ruminants (mediated by microorganisms in the rumen and capable of breaking down cellulosic phytomass indigestible by other mammalian species) is by far the largest source of livestock methane (CH_4) followed by the emissions from manure management (Crutzen et al. 1986; IPCC 2006). Large ruminants convert 5–6% of their feed energy to methane while the rate for pigs is less than 1%, but calculating actual enteric CH_4 emissions on global or continental scales requires assuming averages for the mass of ruminant animals and their prevailing feed quantity, quality and feed digestibility. All of these are highly variable determinants of the fermentation process, and, as a result, the annual emission rates per head of dairy and beef cattle in North America can be more than twice those in India. Similarly, temperature-dependent annual CH_4 emissions from beef cattle manure stored in liquid/slurry pits can be less than 10 kg/head in cool climates and more than 25 kg/head in warm settings.

Not surprisingly, the Intergovernmental Panel on Climate Change list of enteric fermentation emission factors is accompanied by a caveat that all estimates have an uncertainty of ±30–50% (Dong et al. 2006); obviously, this uncertainty can result in large errors when applied to hundreds of millions of animals. Steinfeld et al. (2006) put the global CH_4 emissions from enteric fermentation at nearly 86 Mt in 2004, with cattle accounting for about 77% (and dairy cattle for about 18%) and with mixed livestock production responsible for nearly two-thirds of the total and landless operations accounting for only about 1%. To this must be added CH_4 emissions from manure management estimated by Steinfeld et al. (2006) at roughly 18 Mt in 2004. Not surprisingly, the breakdown of their origins is different, with mixed systems dominating (about 70%) but with CAFOs contributing about 25%, largely due to emissions from pig manure storages. In individual terms, pastured animals have higher ruminal acetic production and hence produce more methane, about 0.4 kg/kg of gain compared to 0.14 kg/kg in a feedlot (Capper et al. 2011).

Nitrous oxide (N_2O) originates from fertilization of soils (no matter if with manures, recycled cop residues or synthetic compounds) used to produce animal feed as well as from denitrification of nitrogen-rich animal manures. Because N_2O fluxes from soils can range over two orders of magnitude, it is particularly difficult to choose a typical rate, and using a single default emission factor for the gas released from nitrogen fertilization – as do Steinfeld et al. (2006) at $0.0125 kg N_2O–N/kg N$ – is yet another source of large uncertainty. Steinfeld et al. (2006) estimated total of N_2O evolved from animal excreta was 3.69 Mt N in 2004.

Sources of carbon dioxide (CO_2) are most diverse: the gas is released during the conversion of forests and wetlands to pastures; from the tillage of soil containing high concentration of organic carbon; from the cultivation of feed crops (directly as fuels to power field machinery, indirectly as fossil energies embodied in inputs of machinery and agro-chemicals, above all inorganic fertilizers whose extraction, synthesis and formulation consumes substantial amount of fossil carbon); from energies required for the preparation of compound feeds that is often preceded by long-distance transportation of their ingredients; from housing, watering and feeding of animals; from their slaughtering and processing; and, finally, from the distribution of meat that now increasingly includes intercontinental transfers of frozen and chilled products.

The first process, conversion of forests to pastures and cropland to be used for cultivating feed crops, is by far the largest source of CO_2 emissions attributable to meat production, and a large error inherent in its estimation is greater than all other, no matter how assiduously accounted for, carbon sources. Steinfeld et al. (2006) assumed that some 3 Mha of forests (averaging 194 t C/ha of phytomass) are converted every year, and that virtually all of that carbon will be rapidly oxidized. These assumptions yielded annual emissions of about 580 Mt of carbon (as CO_2), and all other carbon emissions attributable to livestock – from soils used to cultivate feed, from the burning of fossil fuels used in the synthesis of nitrogen fertilizers, from fossil fuels used on farms (for housing the animals, heating, cooling, lighting and feed preparation and distribution), and from meat processing and transport – are most likely less than 15% of that total. That is well below the minimal value of that estimate's uncertainty: just consider that the tropical rain forest can store anywhere between about 50 and 700 t C/ha (Saatchi et al. 2011).

The standard way to aggregate all of these releases is to express them in terms of overall GWP with the CO_2 equivalent as the common denominator; this conversion takes into account that over a given period of time (25, 50 and 100 years spans are usually chosen), both CH_4 and N_2O are, mole per mole, more potent greenhouse gases than CO_2: over a century, CH_4 rates 21 CO_2 equivalents while N_2O has an order of magnitude higher multiple of 310 (UNFCCC 2012). Steinfeld et al.'s (2006) estimate for 2004 of 2.7 Gt CO_2 equivalent from livestock-related carbon release was nearly 10% of all CO_2 emissions, 2.2 Gt from CH_4 added up to 35–40% of all released methane and 2.2 Gt from N_2O accounted for nearly two-thirds of all anthropogenic emissions of that greenhouse gas. The grand total of 7.1 Gt of CO_2 equivalent (with about 70% from extensive livestock practices) was about 18% of total anthropogenic emissions of greenhouses gases in 2004, a surprisingly high share (higher than the contribution from all forms of global transportation) that received some critical appraisal in research literature as well as a great deal of media attention.

Some of this criticism was inexplicably wrong. Perhaps most notably, Cattlemen's Beef Board (2009) pointed out that Steinfeld et al.'s (2006) study "vastly overestimates" the amount of nitrogenous fertilizers used to produce feed grain for America's livestock (it assumed that 51% of all nitrogen were applied to feed crops and pastures), and that using USDA's data for land planted to feed grain crops and for typical nitrogen applications rates yields the aggregate of no more than 690,000 t that would have been responsible for about 1.75 Mt of CO_2 rather than 11.7 Mt claimed by the FAO. But USDA's nitrogen fertilizer data clearly indicate that in 2010 corn alone received 46% of all applied nitrogen (USDA 2012c).

On the other hand, Goodland and Anhang (2009) argued that CO_2 from livestock respiration is an overlooked source of the gas and that overall greenhouse gas emissions attributable to livestock should be much higher, accounting for no less than half of the total. Unfortunately, they accepted some faulty calculations by Calverd (2005) who concluded that farm animals generate about 21% of all CO_2 attributable to human activities. They claimed that his number "is the only original estimate of its type" (Goodland and Anhang 2009, 12), and using that share they put the total of livestock-derived CO_2 at about 8.8 Gt/year. Two years earlier, Prairie and Duarte (2007) calculated the global respiration of domestic animals at 5.5 Gt CO_2, another wrong total: although they based their calculations correctly on average metabolic rates and used the FAO's total livestock numbers, they assumed highly exaggerated average body masses of animals (cattle at 891 kg, pigs at 200 kg and goats at 89 kg).

The correct respiration rates to use – calculated from first principles (relating metabolism to average body weights) or using numbers carefully established by extensive experiments (Jentsch et al. 2009) – are about 1 kg CO_2/day for pigs (the same value is a good average for the human population, with active adults respiring about 1.5 kg/day), 0.5 kg for sheep and goats and 4 kg for cattle. These rates and the FAO's 2010 livestock numbers yield annual global respiration of about 3 Gt CO_2 or roughly 10% of the gas released in 2010 from the combustion of fossil fuels. In any case, CO_2 emissions from livestock metabolism are not, much like those from human respiration, considered as anthropogenic sources of a greenhouse gas. They result from the metabolism of phytomass that absorbed the gas from the atmosphere and that will again sequester it once it will have been respired and are thus a part of relatively rapid sub-cycle of the global carbon cycle that does not result in any significant net addition to atmospheric CO_2 burden.

In fact, as the total zoomass of domesticated animals has been increasing, it could be argued that these animals represent a small, but rising, carbon sink. As already explained, the carbon stock of domestic animals had more than tripled during the 20th century (from about 35 to about 120 Mt C, with the stock in cattle nearly quadrupling to 90 Mt C). But this is a very small increment compared to losses in deforestation (or additional storage in more productive forests), and, in any case, its effect is easily negated by higher CH_4 emissions. Similarly, CO_2 liberated by regular burning of tropical grasslands (on the order of 2 Gt C/year) in order to prevent the establishment of woody phytomass should not be counted among livestock-related greenhouse gas emissions because the gas is promptly sequestered as the grasses regrow the very next season. But this extensive burning has other undesirable environmental effects, including emissions of nitrogen oxides and hydrocarbons and, above all, the generation of aerosols whose high concentrations affect incoming solar radiation and precipitation in the Amazon (Pöschl et al. 2010).

A more relevant criticism of the FAO's 18% figure is that such a large part of it derives from tropical deforestation in South America, and hence the national shares of livestock-related greenhouse gas emissions are substantially lower in Europe and North America even though those continents support very large numbers of domestic animals. Pitesky et al. (2009) pointed out that Hockstad and Weitz (2009) set the share of live-stock-related US emissions at just 2.8% and that an identical share was found by a 2004 inventory in California. Different analytical boundaries (both of these analyses considered only direct emissions from enteric fer-mentation and manure management) and much higher non-agricultural emissions in the US compared to the global mean explain the difference.

But even if the boundaries matched the FAO analysis, the share would be still much lower as the largest variable, livestock-driven deforestation, is entirely absent in the US where the carbon-absorbing capacity of forests has been actually increasing. The same is true about Europe, and hence any reference to Steinfeld et al.'s (2006) 18% global share of GWP attributable to livestock should always be accompanied by a caveat, namely, that a third of this fraction is due to highly uncertain estimates of grazing-driven deforestation, a process that has nothing to do with modern beef production in affluent countries and that could be much reduced by measures focusing on higher intensity of production rather than on more extensive grazing.

Another difficulty arises due to the conversion of greenhouse gases to a common denominator equivalent to the warming effect of CO_2. Calculations can use different time spans for the GWP (20, 50, 100, 500 years), and the values for such a relatively short-lived gas as CH_4 (average atmospheric residence time of about 12 years) will be much higher for short periods than over longer time spans. The United Nations Framework Convention on Climate Change uses 58 for the 20-year GWP of CH_4 and 21 for the 100-year value and 6.5 for the 500-year equivalent (UNFCCC 2012). This means that methane is disproportionately important in the short term and cutting the numbers of ruminant livestock would bring relatively large immediate reductions in GWP, but that the effect is more than halved in a century.

These realities should be kept in mind when comparing a number of analyses that calculated the carbon footprint (GWP) of different meat production systems. As expected, estimates for beef have been most common (Subak 1999; Ogino et al. 2004; Verge et al. 2008; Peters et al. 2010; de Vries and de Boer 2010). Published extremes range from 5.9 to 36.4 kg of CO_2 equivalent/kg of body weight gain, with most rates between 8 and 16 kg (Flachowsky 2011). When these studies are expressed in kg of CO_2 equivalent/kg of carcass weight, their results range from 8.4 for African to 25.5 for Japanese beef, with the modal rate between 15 and 16 kg, and with relatively small differences between grass-finished and grain-finished meat. For example, for Australian beef, Peters et al. (2010) found the two means at, respectively, 9.9 and 12 kg of CO_2 equivalent/kg of carcass weight. And there are also relatively small differences between conventional and organic production practices, with some studies finding a slightly higher and others slightly lower carbon intensity for the latter (Flachowsky 2011).

A detailed European assessment put the total greenhouse gas emissions at 28.7 kg of CO_2 equivalent for per kilogram of beef, with the rates for pork and poultry at, respectively, 11.2 and 3.6 kg (Weidema et al. 2008), while Flachowsky (2011) concluded that under European conditions, average CO_2 emissions are 0.12 kg/kg of dry matter of roughage and 0.22 kg/kg of concentrate feeds and, for comparison, found carbon footprint of milk production at mostly between 0.8 and 2 kg of CO_2 equivalent/kg. As for the CO_2 emissions alone, Flachowsky and Hachenberg (2009) found the published ranges (kg/kg of product) between 0.1 and 0.2 for pasture, hay and silage and between 0.2 and 0.5 for feed grains; for meat, the ranges are between 8 and 18 kg for beef, 3 and 6 kg for pork and 2 and 6 kg for poultry, which means that the releases for edible portions may be easily around 25 kg for beef. Peters et al. (2010) calculated a wider range for Australian beef, ranging from about 6 to 18 kg CO_2/kg of carcass weight.

Finally, an interesting finding that concerns the relative emissions from production and transportation. Weber and Matthews (2008) found that even in the US with a considerable transportation component (with average food delivery of 1,640 km and life-cycle supply chain mean of 6,760 km), greenhouse gas emissions along the meat chain are dominated by production (83% of the total), and only 11% of all life-cycle emissions can be attributed to transportation (the rest going to final delivery and retail).

Ruminants have a unique ability to convert phytomass that is not digestible by other vertebrates into high-quality protein food: rational meat production should maximize this option. Photo by the USDA.

5

Possible Futures

Where does this all leave us? There is no doubt that human evolution
has been linked to meat in many fundamental ways. Our digestive tract
is not one of obligatory herbivores; our enzymes evolved to digest meat
whose consumption aided higher encephalization and better physical
growth. Cooperative hunting promoted the development of language
and socialization; the evolution of Old World societies was, to a
significant extent, based on domestication of animals; in traditional
societies, meat eating, more than the consumption of any other category of foodstuffs, has led to fascinating preferences, bans and diverse
foodways; and modern Western agricultures are obviously heavily meat-
oriented. In nutritional terms, the links range from satiety afforded by
eating fatty megaherbivores to meat as a prestige food throughout
the millennia of preindustrial history to high-quality protein supplied
by mass-scale production of red meat and poultry in affluent
economies.

But is it possible to come up with a comprehensive appraisal in order
to contrast the positive effects of meat consumption with the negative
consequences of meat production and to answer a simple question: are
the benefits (health and otherwise) of eating meat greater than the
undesirable cost, multitude of environmental burdens in particular, of
producing it? Difficulties in answering these questions are immediately

Should We Eat Meat?: Evolution and Consequences of Modern Carnivory,
First Edition. Vaclav Smil.
© 2013 John Wiley & Sons, Ltd. Published 2013 by John Wiley & Sons, Ltd.

obvious: as useful and as revealing such a comparison might be, it deals in contrasting quantities that have few, or no, common denominators and an enormous range of qualities that cannot be readily (or at all) quantified.

These qualities range from the effects of protein deficiency on developing brains to cruelty to animals, and from both positive and negative consequences of moderate meat consumption by adults to lingering burdens of livestock-generated gases in the biosphere. Even so, before turning to appraisals of futures with more or less meat or entirely without meat, I will present a brief summary of a still accumulating understanding of major positive and negative effects and specific links with major civilizational diseases and human longevity. Once that is done, I will turn to appraisals of alternatives to modern meat eating and their chances of near-term success.

Killing animals and eating meat have been significant components of human evolution that had a synergistic relationship with other key attributes that have made us human, with larger brains, smaller guts, bipedalism and language. Larger brains benefited from consuming high-quality proteins in meat-containing diets, and, in turn, hunting and killing of large animals, butchering of carcasses and sharing of meat have inevitably contributed to the evolution of human intelligence in general and to the development of language and of capacities for planning, cooperation and socializing in particular. Even if the trade-off between smaller guts and larger brains has not been as strong as is claimed by the expensive-tissue hypothesis, there is no doubt that the human digestive tract has clearly evolved for omnivory, not for purely plant-based diets. And the role of scavenging, and later hunting, in the evolution of bipedalism and the mastery of endurance running cannot be underestimated, and neither can the impact of planned, coordinated hunting on non-verbal communication and the evolution of language.

Homo sapiens is thus a perfect example of an omnivorous species with a high degree of natural preferences for meat consumption, and only later environmental constraints (need to support relatively high densities of population by progressively more intensive versions of sedentary cropping) accompanied by cultural adaptations (meat-eating restrictions and taboos, usually embedded in religious commandments) have turned meat into a relatively rare foodstuff for majorities of populations (but not for their rulers) in traditional agricultural societies. Return to more frequent meat eating has been a key component of a worldwide dietary transition that began in Europe and North America with accelerating industrialization and urbanization during the latter half of the 19th century. In affluent economies, this transition was accomplished during the post-WW II decades, at a time when it began to unfold, often very rapidly, in modernizing countries of Asia and Latin America.

As a result, global meat production rose from less than 50 Mt in 1950 to about 110 Mt in 1975; it doubled during the next 25 years, and by 2010 it was about 275 Mt, prorating to some 40 kg/capita, with the highest levels (in the US, Spain and Brazil) in excess of 100 kg/capita. This increased demand was met by a combination of expanded traditional meat production in mixed farming operations (above all in the EU and China), extensive conversion of tropical forests to new pastures (Brazil being the leader) and the rise of concentrated animal feeding facilities (for beef mostly in North America, for pork and chicken in all densely populated countries).

This, in turn, led to a rise of modern mass-scale feed industry that relies primarily on grains (mainly corn) and legumes (with soybeans dominant, fed as a meal after expressing edible oil) combined with tubers, food processing residues and many additives to produce a variety of balanced feedstuffs containing optimal shares of carbohydrates, proteins, lipids and micronutrients (and added antibiotics). But it has also led to a widespread adoption of practices that create unnatural and stressful conditions for animals and that have greatly impaired their welfare even as they raised their productivity to unprecedented levels (with broilers ready for slaughter in just six to seven weeks and pigs killed less than six months after weaning).

Meat is undoubtedly an environmentally expensive food. Large animals have inherently low efficiency of converting feed to muscle, and only modern broilers can be produced with less than two units of feed per unit of meat. This translates into relatively large demands for cropland (to grow concentrates and forages), water, fertilizers and other agrochemicals, and other major environmental impacts are created by gaseous emissions from livestock and its wastes; water pollution (above all nitrates) from fertilizers and manure is also a major factor in the intensifying human interference in the global nitrogen cycle.

But to say that "1 kg of fresh vegetables is environmentally preferable to 1 kg of meat" (Reijnders and Soret 2003, 666S) – a conclusion based on a Swiss study (Jungbluth et al. 2000) according to which organic vegetables incurred 0.016 ecopoints/kg compared to 0.08 ecopoints/kg for meat – is a disingenuous comparison that may reinforce anti-meat bias but that completely leaves out the qualitative differences between the two categories of foodstuffs. Moreover, balanced nutrition, especially that of young children, cannot be achieved by eating only organic vegetables or only red meat, and merits of both of those food categories should be compared only within a broader dietary framework.

Similarly, when Kingsolver (2002) writes that 0.25 lb hamburger requires 100 gallons of water, a cup of gasoline, 1.2 lb grain and 1.25 lb topsoil lost to erosion, an uncritical reader may be impressed by a combination of these high environmental burdens, but a critical examination must (as already explained) distinguish among different categories of those water "requirements" and should provide at least some comparisons with energy, feed and erosion impacts of alternatives that would produce nutrition of a comparable quality. Most importantly, meat's environmental burdens can be substantially reduced by a combination of increased efficiencies and moderate average per capita meat intakes: rational meat eating is definitely a viable option.

Opportunities for higher efficiency can be found all along the meat production–consumption chain. Agronomic improvements – above all reduced tillage and varieties of precision cropping (including optimized irrigation) – can reduce both the overall demand for natural resources and energy inputs required for feed production while, at the same time, improving yields, reducing soil erosion, increasing biodiversity and minimizing nitrogen leakage (Merrington et al. 2002). Many improvements can lower energy used in livestock operations (Nguyen et al. 2010), reduce the specific consumption of feed (Reynolds et al. 2011) and minimize environmental impacts of large landless livestock facilities (IST 2002). Considerable energy savings can also be realized by using better slaughter and meat processing methods (Fritzson and Berntsson 2006).

And there is nothing inevitable about the further spread of antibiotic-resistant bacteria due to the overuse of antibacterial compounds in meat production. Unlike the US, the EU banned the use of antibiotics as livestock growth promoters in 2006. The ban included monensin sodium used in fattening cattle, salinomycin sodium for piglets and pigs, and avilamycin and flavophospholipol used for pigs, poultry and cattle (EC 2005). Perhaps most impressively, Denmark, the world's leading exporter of pork, has demonstrated how general antimicrobial use in pigs can be reduced by as much as 60% with concomitant production increases (Aarestrup 2012). Similarly, reduced use of antibiotics in poultry had no negative effects either on the total productivity per unit of floor area or the amount of consumed feed (Emborg et al. 2001).

Alternatives to high levels of meat consumption range from more widespread vegetarianism to higher intakes of other animal foodstuffs supplying high-quality protein. What, then, is the likelihood that not just only some concerned individuals but modern societies at large will choose some of these alternatives or that they will, at least, start moving in those directions? Are there any realistic prospects for a voluntary, deliberate, mass-scale and lasting retreat from meat consumption in affluent countries? Are there any indications that the populations of populous, rapidly modernizing countries of Asia, Latin America and richer parts of Africa will depart from a worldwide pattern of dietary transition and will not want to spend part of their rising disposable income on eating more meat?

Meat consumption in modernizing countries is highly dependent on disposable income, but the post-2008 realities make it clear that any long-range forecasts of economic performance (and hence of the most likely corresponding average meat intakes) are highly questionable. While the levels of future income-driven meat consumption in Asia or Africa remain uncertain, it should be possible to lower average meat intakes in affluent countries and transform the entire population into moderate carnivores. This is a realistic option as many high-income countries have already seen decades-long saturation in average meat intakes, and as quite a few have experienced noticeable declines of total consumption, and even more so of particular meat varieties. Meat price elasticities make it clear that dearer meat in general, and more expensive beef in particular, would lower both overall and specific demand much faster than the slow unfolding of dietary preferences driven by aging and health concerns.

Toward Rational Meat Eating: Alternatives and Adjustments

Many excesses that have come to characterize modern, intensive and mass-scale meat production – particularly those concerning inconsiderate treatment of animals, their stressful transportation and often callous slaughtering, as well as the narrow concentration on rapid weight gains obtained through confined feeding and a heavy use of antibiotics – should be criticized, and they should be eventually either banned or substantially modified. Similarly, the extent to which natural ecosystems (particularly tropical forests) have been (and are still being) converted to pastures as well as frequent overgrazing of grasslands and damage done by livestock to soils and the impact of atmospheric emissions and water pollution generated by livestock are matters of considerable concern that require appropriate modifications and improvements. And we should question the extent to which arable land is devoted to monocultures (mostly corn and soybeans) whose harvests are needed for animal feeding.

But the prevalence of these objectionable practices and the validity of these concerns are not convincing arguments against meat eating. Those practices are not inherent prerequisites of large-scale meat production; they are essentially malpractices committed as a part of a short-sighted quest for maximized meat output at minimized cost. Our understanding of livestock requirements, feed production and animal feeding, slaughtering and processing makes it possible to practice balanced and rational ways of meat production aimed at minimizing its environmental impacts and maximizing its health benefits.

There is nothing foreordained about today's dominant ways of meat production, and many substantial, highly positive changes can be effected with some fairly simple adjustments. We could produce globally several hundred millions of tons of meat without ever-larger confined animal feeding operations (CAFOs), without turning any herbivores into cannibalistic carnivores, without devoting large shares of arable land to monocropping that produces animal feed and without subjecting many grasslands to damaging overgrazing – and a single hamburger patty does not have to contain meat from several countries, not just from several cows. And there is definitely nothing desirable to aim for ever higher meat intakes: we could secure adequate meat supply for all of today's humanity with production methods whose energy and feed costs and whose environmental impacts would be only a fraction of today's consequences.

Uncertainties regarding the future rates of global population and economic growth and the environmental change (particularly the effects of global warming on the productivity of pastures and feed crops) make all forecasts of future meat demand and production no better than educated guesses – but unless we are prepared to tolerate even greater environmental impacts, the need for substantial changes of today's meat-producing practices is clear even if the prospects were for only a moderate expansion of average per capita meat consumption. Opportunities for improved efficiencies along the entire meat production chain abound, starting with better ways of cultivating feed crops (increasing average yields, moving away from monocropping to rotations with leguminous forages, adopting reduced tillage, using less wasteful irrigation and fertilization methods) and ending with substantial energy savings (and hence lower greenhouse gas emissions) through better housing of animals and improved management of manures.

As in the past, economic imperatives and environmental regulations will keep on diffusing adoption of these more efficient ways, but they alone would not suffice to bring about a more fundamental transformation: that will require not only making the existing system more efficient (and hence less environmentally intrusive) but also aiming at restrained meat consumption that can be achieved by abstaining from meat, by choosing alternatives to meat protein, and by gradually (but substantially) reducing typical meat intakes in affluent nations and keeping the supply at moderate levels in modernizing countries. Specialized publications in agronomy, plant science and nutrition, range science, animal feeding and manure management abound with discussions of improvements in these fields, but examinations of the other long-term strategies – meatless diets, alternatives to meat protein and moderation of meat consumption – have been much less common, and that is why I will examine them more closely, and in the indicated order, in this closing chapter.

I will demonstrate a surprisingly large potential for what I call desirable meat production, for the ways of raising livestock that combine reduced environmental impacts with improved animal welfare and with optimized health benefits. This, I believe, is by far the best long-term choice in the quest to base modern meat eating on rational foundations. As I always do in my writings, I will not offer any time- and quantity-specific forecasts; instead, I will close the book with some probabilistic musings about short- (a decade) and mid-term (a generation, or 20–25 years) prospects for the global meat output and its composition and about the most likely changes in global and national per capita meat consumption. Meat consumption is a part of our evolutionary heritage; meat production has been a major component of modern food systems; carnivory should remain, within limits, an important component of a civilization that finally must learn how to maintain the integrity of its only biosphere.

Meatless diets

I start with what I think is perhaps the least promising option for a major transformation of today's meat consumption habits, with the assessment of meatless diets. Certainly, the most radical but the least practical way to reduce meat, or any food, consumption is to reduce food consumption through imagination. The claim to the efficacy of this option rests on just five experiments, each one with only 51–81 participants: their conclusion was that people who repeatedly imagined eating a particular food many times did subsequently eat less of it than those who imagined eating that food fewer times, thought about eating a different foodstuff or did not imagine eating anything (Morewedge et al. 2010). Clearly, this option is not going to be a likely means of population-wide reduction in the demand for steaks, hamburgers and fried chicken.

A more realistic choice – but one that proves, on a closer examination, still rather daunting, besides being, most likely, not as efficacious as it has been claimed – is to practice deliberate restraint in food intake, an approach that is now usually put into a category of caloric restriction (CR) whose greatest benefit is to extend the average lifespan of any heterotrophic organism (Mattson 2005). This effect was discovered first with rat experiments during the 1930s (McCay et al. 1935), and it is achievable only when adequate levels of all nutrients are combined with reduced energy intakes (hence calorie restriction with optimum nutrition). Its benefits include postponed onset of major illnesses, reduced blood pressure, improved glucose metabolism, enhanced immune response and greater longevity, and it has been demonstrated in a wide range of species ranging from protozoa to mammals (Weindruch and Walford 1988; Sohal and Weindruch 1996).

Wilcox et al. (2007, 173) believe (based on their work in Okinawa where the average energy intakes are among the lowest in the world) that CR "not only *will* work but in fact available epidemiological evidence indicates that CR *may already have* contributed to an extension of average and maximum lifespan." But Speakman and Hambly (2007) make several cautionary qualifications. They point out that the results of animal studies conducted in optimal, pathogen-free environments with non-stressed subjects are not applicable to people exposed to infections and required to work; moreover, CR benefits would accrue only with life-long adherence and decline considerably as the age of CR onset is delayed. And perhaps the greatest challenge would be to live a life of constant hunger: monitoring of experimental responses indicates that even prolonged restricted diets do not diminish that feeling and that the animals are continuously hungry: enduring that would be a tall challenge to meet in everyday life.

That is why Le Bourg (2006) concludes that CR would not increase longevity of any species able to leave a place where it is subject to semi-starvation: adopting CR voluntarily on a large scale is unlikely, and any population subjected to it involuntarily will choose to migrate. Phelan and Rose (2006) make perhaps the most fundamental cautionary point: dietary reaction norms connecting longevity to CR rodents are fairly steep in rodents (small changes translate into large gains), but human reaction norms are very different, and any gains from even an assiduously practiced human CR would be minor, about an order of magnitude less than expected in mice. This conclusion was strongly supported by the results of a long-term study reported by Mattison et al. (2012) that found no survival gains in young and older age rhesus monkeys subjected to calorie restriction.

While imaginary eating and restricted diets that leave people permanently hungry are too impractical to make any real difference, vegetarianism does not ask for such sacrifices. Vegetarians can eat *ad libitum*, and they do not even have to abstain from animal foodstuffs other than mammalian and avian meats. That sounds illogical because it leaves three entire categories of animal foodstuffs, dairy products, eggs and all seafood – but (as I have already noted in the preface) the habit's current usage departs from strict logic of defining vegetarianism as food consumption limited to plant foodstuffs (that is now called veganism) and allows for a progressive three-step inclusion of non-meat animal foods: lacto-vegetarians obviously drink milk, can butter their bread, put cream into their coffee and eat yogurt and cheeses with their grains and roots; lacto-ovo vegetarians add eggs to their dairy foods; and cumbersomely called lacto-ovo-pisci vegetarians also eat any animals coming from the sea or from freshwaters – or simply anything but terrestrial meats.

Van der Kooi (2010) dismissed all of these variations and argued that if a vegetarian is motivated by arguments that focus on animals, then the only consistent option is to become a vegan. Obviously, as long as their overall food energy intakes coming from balanced diets are adequate, none of these long-definition vegetarians should have any nutritional deficiencies as they have access to high-quality animal proteins and various sources of micronutrients (Van Winckel et al. 2011); and, as a British national survey showed, "vegetarian" children might actually consume more milk and dairy products than their peers (Thane and Bates 2000). Similarly, true vegetarian diets can also be perfectly adequate for adults: again, as long as overall energy and protein intakes are commensurate with the respective vegan diets, they can support healthy, active lives as well as healthy pregnancies, and in old age do not have an adverse effect on bone loss and fracture (Mangels 2008; Ho-Pham et al. 2012).

With infants and children on solely plant diets, it is trickier: without proper planning, they could experience restricted energy intake, impaired growth and micronutrient deficiencies, above all of vitamin B11, iron and zinc that could cause serious mental and cardiac problems (Codazzi et al. 2005; Giannini et al. 2006). Given the lower digestibility of plant proteins, it is highly advisable to increase vegan protein intakes by 30–35% for infants younger than two years, 20–30% for children up to six years and 15–20% for older children, and attention should also be paid to an adequate supply of foods containing fatty acids. With proper meal planning and with suitable micronutrient supplements (B12, D, iron and zinc may be needed for older infants and young children), all possible dangers can be avoided and vegan children could grow as is normally expected (Mangels and Messina 2001; Antony 2003; Amit 2010).

These realities and assurances have been known for decades, and the promoters of vegetarian diets have also stressed the environmental, planet-sparing benefits of meatless diets. Indeed, a best seller of this *genre* that combined the counter-culture ethos of the 1960s with the new environmental awareness of the early 1970s, *Diet for a Small Planet* promised "delicious food combinations of protein-rich meals without meat" (Lappé 1971). A few years later, in *The Book of Tofu*, Shurtleff and Aoyagi warned about a serious food crisis and predicted that within 10 to 20 years, the sources of America's dietary protein will be completely reversed, with 80% originating in plant foods, and that tofu shops will spread around the country and "make an invaluable contribution to the betterment of life on our planet" (Shurtleff and Aoyagi 1975, 13).

Three decades after her first book, Lappé returned with a variation on the same topic, written with her daughter Anna, that contains such recipes as parsnip patties, carrot salad with hot goat cheese, and marinated tofu over roasted Yukon gold *caponata* with sun-dried tomato and basil *coulis* (Lappé and Lappé 2002). Tasty if well prepared, well meaning, undoubtedly, but even the first suggestion is a bit precious. As for the second one, how many Americans are habitually preparing hot French goat cheeses? And the last choice reads as an unintended caricature of *nouvelle cuisine* assemblages (how many Americans even know about *caponata* and *coulis*?). The book's title is *Hope's Edge*, but I doubt that this edge will be parting the way for new American eating in the first half of the 21st century.

Hardly any parting has been done since the counter-cultural appeals of the 1970s: true, average per capita meat consumption has reached a plateau, but there has not been any sign of precipitous decline; animal foods still supply about 65% of America's dietary protein, and small tofu shops never took root in America's malls (and they have been rapidly disappearing in Japan). And although tofu (as well as a variety of tofu-based meat substitutes) is now always carried by all major US supermarket chains, its widespread availability has not made any noticeable difference to America's diet: annual per capita consumption remains minuscule, rising higher only among vegetarians. On the other hand, most of the popular competitive eating contests (truly revolting all-American spectacles) involve eating meat as hot dogs or hamburgers (Fagone 2006), and in order to make excessive meat consumption even more effortless, in April 2012 Pizza Hut launched a "limited edition" of hot-dog-stuffed crust pie.

And even America's constant dieting craze has done little to convince significant shares of meat-eating populations to abstain from meat, or even from all animal foodstuffs: studies show that prevalence of all forms of "vegetarianism" is no higher than 2–4% in any Western society and that long-term (at least a decade) or life-long adherence to solely plant-based diets is less than 1%, and that the single largest category are younger, more educated females living in households with higher incomes (Frank and White 1994; The Vegetarian Resource Group 2001). Many of them are "ideological" or "moral" vegans who chose the diet for its presumed superiority (often perceived in quasi-religious terms) rather than pragmatic vegetarians who switched to a meat-free diet as a means of improving their health after discovering that they have an elevated risk of cardiovascular disease (CVD) or cancer mortality (Ginter 2008).

And one of the largest groups among vegetarians consists of teenage females, and Worsley and Skrzypiec (1997) found that those who make the choice are more concerned about their appearance than their peers and report a greater frequency of extreme weight loss behaviors. This suggests a link between teenage vegetarianism and female adolescent identity rather than a fundamental concern about animal welfare or environmental impacts of eating meat. But the choice will do no harm for adults; in fact, the vegetarian imperative (Saxena 2011) has been largely justified by the health benefits of the diet that have been repeatedly demonstrated by reduced cholesterol intakes and slightly lower blood pressure among populations abstaining from meat.

But does vegetarianism really translate into notable quality-of-life and longevity gains; does it really improve the chances for a longer life, particularly when compared with moderate meat eating? The best answer is that it helps to change the pattern of morbidity and mortality, but that it has a very limited effect on longevity as vegetarians living in the same overall settings (with access to the same preventive and therapeutic health care and with similar housing condition) do not live significantly longer than moderate meat eaters. Study of Seventh-day Adventists offered the first widely publicized statistical proof of benefits from following a vegetarian diet: even though only half of male Adventists were vegetarians, the group's CVD mortality was lower than in a control American population, and the diet also appeared to be protective against colorectal cancer (Kahn et al. 1984).

But rather than looking at specific diseases, it is much more revealing to note the final outcome: the world's longest living population is in Japan, a country whose per capita food supply now averages more than 50 g/day of animal protein, with about 40% coming from seafood and 30% from meat (Smil and Kobayashi 2012). And Europe's three societies with the highest life expectancy – Sweden, Norway and Iceland – have diets with substantial quantities of meat and a large amount of dairy products (Ginter 2008). These nationwide realities have been recently confirmed by a large-scale study that compared the mortality of British vegetarians with a matched omnivorous population (but one with moderate meat intake, averaging close to 25 kg/year for men).

In the European Prospective Investigation into Cancer and Nutrition, which began during the late 1990s with 65,000 participants, mortality from circulatory diseases and from all causes was not significantly different between meat eaters and vegetarians, and the only caveat the authors could include was that the study was "not large enough to exclude small or moderate differences for specific causes of death" (Key et al. 2009b, 1613S). That is hardly a strong advertisement for veganism.

And the difference in environmental impacts does not have to be huge: meaty Western diets (at least 3,500 kcal/day, about 30% of food energy from animal products and about 10% from meat) claim (depending on the share of meat in general and beef in particular) up to 4,000 m² of farm-land/capita compared to as little as 700–800 m²/capita for a vegetarian diet produced by high-intensity cropping – but a diet containing about 15% of food energy from animal foodstuffs and supplying adequate average of 2,000 kcal/day needs no more than 1,100 m²/capita. In any case, voluntary population-wide abstention from eating animal foodstuffs is extremely unlikely.

Meat substitutes and cultured meat

Meat substitutes (mock or *faux* meats) have a long history as vegetarian Asian cuisines (Hindu, Chinese and Japanese) developed dishes imitating not just the flavor and texture of meat but often recreating meticulously its appearance; these creations were based on processed soybean products (*tofu* and *tempeh*), wheat gluten, mushrooms and pulses, and so are their modern commercial meat substitutes that are available in a variety of fresh, chilled or frozen forms and that often contained specific flavorings (Shurtleff and Aoyagi 2004). This, of course, begs a question why vegetarians, the benefactors of a superior diet, would have any need to eat fake meat products – but clearly many of them crave fake cheeseburgers, wieners and a Thanksgiving turkey.

Although People for Ethical Treatment of Animals (PETA) claims that "a terrific array of vegetarian mock meats already exists" (PETA 2012), new varieties keep appearing. UK-based Quorn™ is now the world's largest seller of frozen meat substitutes not based on soy products but on single-cell mycoprotein from *Fusarium venenatum* grown in vats (Quorn 2012). Fungi, as already explained, have high-quality proteins, and so Quorn's vegan burgers and "chicken" patties, cutlets and nuggets may be thought of as reconstituted mushrooms. Quorn's leading British competitor now offers soy- and wheat-protein-based vegetarian breakfast and cocktail sausages, wieners, ground "meat," "chicken" nuggets and drumsticks as well as Wienerschnitzel (Tivall 2012).

Leading American producers are Turtle Island Foods, Lightlife Foods, Boca Foods and Beyond Meat™. Turtle Island Foods are the makers of irresistibly named Tofurky delights that include not just "a bird-free feast, but with all the flavor and trimmings you've always enjoyed" but also Tofurky *kielbasa*, *chorizo* and jerky, and tempeh bacon. Lightlife Foods, part of America's top giant agribusiness ConAgra, offers frozen soy-based Asian Sesame Chik'n and Olé Santa Fe Chik'n. One commentator believes that Beyond Meat™ is producing "fake meat so good it will freak you out" (Manjoo 2012), and the company boasts that its Chicken-Free Strips "have all of the convenience, taste, and tenderness you expect from real chicken – without the bad stuff (no saturated or *trans* fat, no cholesterol, no gluten, no antibiotics, no GMOs … and no meat)" – as long as you prefer their natural vegan chicken flavor and added titanium dioxide as white food coloring (Beyond Meat™ 2012).

US sales of these meat substitutes were up by 10% in 2011 (Chaker 2012), but at about $270 million that was still equal to less than 0.2% of the country's annual meat sales that reached about $160 billion in 2010 (AMI 2012). Obviously, overall acceptance of these overt meat substitutes remains marginal. Hoek et al. (2011) found that unfamiliarity with these products and their lower sensory attractiveness and common food neophobia were the key determinants of their poor general acceptance. Even after the much publicized meat-safety crises (European bovine spongiform encephalopathy above all) that scared many European meat consumers, the market share of these substitute products as a component of Dutch meals (by volume) remained just 1% (Hoek et al. 2004). That is why I have concluded that the substitutes would have a more important role when used covertly as additives or extenders in ground and processed meats (Smil 2002).

Although there are no global meat consumption data disaggregated by final use categories, I estimated that (based on a variety of national statistics and known dietary preferences) in the year 2000 ground meats added up to at least 15 Mt and that as much as 20 Mt were used in processed meat products. Adjusted for 2010, this means that at least 45 Mt of meat were eaten in the forms where additions of plant proteins would be both practical and would not be objectionable as it would go largely unnoticed, in some cases being just an extension of existing practices. In sausage production, the USDA already allows additions of up to 3.5% of cereal and soy flour or soy concentrate and of up to 2% of soy protein isolates. Adding 5% of plant protein to all processed meat would displace at least 1 Mt of meat, but that is only an equivalent of annual meat production of Austria.

Potential for partial substitution by extending the lean content of ground beef could be even greater: assuming that its annual global consumption is about 20 Mt, replacing just 5% of that total by plant protein would save production of 1 Mt of beef, and with better ways of adjusting for texture and taste, that share could be eventually increased to 10%, displacing 2 Mt of beef every year. If protein substitutes could eventually supply 25% of the nutrient present in 2010 in ground meat and in processed meat products, that substitution would result in net savings (after accounting for the mass of plant protein that would have to be produced or modified as a substitute) of about 75 Mt of concentrated feed, or on the order of 8% of worldwide concentrate feed production.

These would be significant environmental savings, but benefits an order of magnitude greater could come only from a radically new way of meat production: not by substitution using plant- or mushroom-derived proteins but by using animal tissues to culture meat on an industrial scale. Besides doing away with large-scale mistreatment of animals and with inefficient animal feeding that leads to extensive land, atmospheric and aquatic environmental impact, cultured meat's other advantages would include a major reduction of long-distance meat shipments, significantly lowered risk of meat-borne diseases and a possibility to adjust the ratio of saturated-to-polyunsaturated fatty acids.

The idea of *in vitro* meat has been around for decades as a part of broader expectation for perfectly nutritious synthetic food. Ford (2011) chronicles the first research efforts that began during the late 1990s in the Netherlands where the work, now proceeding in several centers (Amsterdam, Eindhoven, Utrecht, Wageningen), was boosted in 2007 by a government investment of €2 million. The first international *in vitro* meat symposium was held in April 2008 in Norway. In the same year, PETA made a major departure from its hostility to consuming any meat products and offered a $1 million reward to anybody able to produce, and bring to market, *in vitro* meat (PETA 2008).

The contest specifications set some very demanding targets: to produce *in vitro* chicken meat whose taste and texture would be indistinguishable from the real flesh when tasted both by vegetarians and carnivores; final taste and texture is to be evaluated by a panel of PETA judges using the organization's own fried "chicken" recipe and the meal must score at least 80 points to win, to manufacture this acceptable product in quantities sufficient to be sold commercially at competitive prices in at least ten US states and to do this all by June 30, 2012. These conditions guaranteed a failure: prize offered to stimulate an invention is normally satisfied with the demonstration of a breakthrough and does not set goals for mass marketing which, particularly in this, will depend heavily on the pace and complications of an inevitable process of government review that, given the nature of an eventually successful synthesis, will not be rather lengthy and will be further slowed down by interventions of many interested parties. Engber (2008) was thus right when he called PETA's reward "the bogus $1 million meat prize."

Very slow progress in developing *in vitro* meat is obviously attributable to the challenge of replicating the complexity of animal muscles and related tissues. Replicating a skeletal muscle with its connective, enervating and blood-supplying tissues is a forbidding challenge that becomes even greater with red meats that should contain at least a minimum of marbling, without which they would not have expected taste when roasted. Not surprisingly, that is why white chicken meat (essentially pure protein) has been taken up first as a relatively easier challenge. There are two basic approaches to *in vitro* meat production: cultivating muscle cells or producing complete muscles (Edelman et al. 2005; Datar and Betti 2010). The first, simpler technique would attach embryonic myoblasts or adult myosatellite cells either on a collagen meshwork scaffold or on collagen beads (because the carrier would have to be edible) suffused with culture medium in a bioreactor.

Diffusional limitations restrict the scaffold-based growth of myocytes to layers just 100–200 μm thick, and even then layering would produce tissue lacking the structure of muscle tissue, and the tissue could substitute only for ground boneless meat. Growth of a fully structured muscle tissue would have to rely on muscle explants in suitable culture media, a challenge made particularly difficult by the absence of blood circulation. With uncertainties about Hayflick limit (number of cell doublings), it is impossible to know how much meat a single cell could eventually produce. Even if these techniques became fairly effective on a small experimental scale, their scaling up is predicated on the availability of an affordable culture media, on the presence of requisite growth factors (produced primarily by liver) and on an eventual development of massive bioreactors (on the order of 1,000 m^3 compared to bench assemblies with volumes of a few liters) to produce commercial amounts of cultured tissues.

If *in vitro* meat were to displace just 10% of recent annual meat production, the processes would have to generate about 30 Mt/year. For comparison, the total production of US antibiotics done by processes using large-scale bioreactors is now on the order of 20,000 t/year – after nearly seven decades of development and expansion (Nawaz et al. 2001). Given these realities, just a convincing demonstration of a viable mid-scale (hundreds of kilograms) process yielding cultured meat of acceptable properties and taste would be an impressive accomplishment. And even when taking a long-term view, Ford (2011) rightly doubts that cultured meat could make livestock farming obsolete.

A sheer complexity of creating such complex structures *in vitro* may mean that mycoproteins might be a better choice for meat-like cultures, but in any case it will take decades before productions could be scaled up – and even that success would not be, given the common food neophobia, a guarantee of eager mass acceptance. Grazing animals will remain indispensable in order to maintain many cherished landscapes: an overwhelming majority of Europe's ubiquitous meadows is entirely artificial, created and maintained by centuries, even millennia, of grazing. This leaves us to explore two other possibilities: the substitution of meat by other sources of animal protein and the extent to which optimal (less meaty) diets could affect the future meat production.

Protein from other animal foodstuffs

Gustatory experience and the nutritional profile of seafood, eggs and dairy products differ from meat in some fundamental ways, and some of those foodstuffs (above all shellfish, cephalopods, milk) are avoided not just by allergy-prone individuals but have been traditionally disdained by entire cultures (particularly by non-milking societies of East Asia). While other animal foodstuffs can never be thought of as truly interchangeable substitutes for meat, they are nutritionally equivalent in terms of protein. As already explained, all proteins of animal origin are of the highest quality, and all are easily digestible; so from a purely nutritional perspective, it would make no difference if those proteins are delivered in meat, milk, eggs or seafood.

And once the complete nutritional compositions of animal foodstuffs are compared, fish is the superior choice: species with white meat are essentially just pure protein, while lipids in fatty fish contain high levels of two ω-3 polyunsaturated fatty acids, eicosapentaenoic acid and docosahexaenoic acid, whose consumption has been associated with reduced inflammation, lower low density lipoprotein cholesterol levels and reduced incidence of coronary heart disease. And while egg yolks contain relatively high amount of cholesterol (480 mg/100 g compared to 80–100 mg/100 g for red and white meats and 40–80 mg/100 g for fish), eating one egg every day contributes an order of magnitude less to an individual's metabolic burden than does the internally generated cholesterol (mainly in liver). Moreover, it is quite easy to separate egg whites from yolks and that way consumers can enjoy the best protein without any cholesterol.

Similarly, it is easy to remove a desired amount of (saturated) fat from milk, and consumers now have a choice between whole milk (typically 3.4% fat) and milks with fat content reduced to 2%, 1% and 0%. And if delivering optimal diets would be as simple as choosing foodstuffs that combine high quality with easy digestibility and relatively low environmental impact, then milk and dairy products would claim the leading position. Milk's moderate environmental impact is mainly due to the fact that its production is by far the most efficient way of transforming cellulosic biomass into a high-quality animal food, and the overall efficiency can be further increased by supplementing forages with concentrates.

As a result, dairy cows fed well-balanced diets need only about 0.8 kg of feed (as an equivalent of feed corn) to produce 1 kg of milk. Century-long US feed efficiency average (that applies, as in the case of previously cited meat conversion efficiencies, to the entire milk-producing herd) shows fluctuations superimposed on a declining trend from about 1.2 in the early 20th century to less than 0.9 one hundred years later (Smil 2001a). Meat-to-milk mass conversion ratio of 0.6 translates (depending on the average energy density of feed and with milk energy at 3 MJ/kg) into gross energy conversion efficiencies of 55–67%, and accumulated cow fat can be converted to milk with efficiency approaching 80%, rates that no feed-to-meat conversion can approach.

Consequently, it is natural to ask – given the protein equivalence of animal foodstuffs, commonly high nutritional and gustatory appreciation of seafood, eggs and dairy products in most societies, and a well-documented possibility of changing dietary habits over time – what shares of today's meat consumption could be realistically displaced by higher intakes of other animal foodstuffs. I will start this inquiry with what appears to be the least promising substitution: would not the efforts to boost dairy consumption *in lieu* of meat run into the metabolic limits due to a well-known fact that many populations have a high incidence of lactose intolerance, a genetic trait caused by low or very low levels of the enzyme that breaks down that milk?

True, lactase deficiency is common in several parts of the world, but it does not stand in the way of increased consumption of dairy products, even in the societies that had no historical tradition of milking. Only a tiny share of genetically disadvantaged babies is born with lactase deficiency, and the rest easily metabolizes lactose in maternal milk. After weaning, intestinal concentrations of lactase begin to decline, and eventually most children in those societies that have not kept milking animals have little of that enzyme left, and the prevalence of lactase deficiency is commonly 70% and sometimes more than 95% among non-milking cultures of sub-Saharan Africa and East Asia. In contrast, in those societies where milking was practiced either by pastoralists (be it in Africa or Asia) or by farmers (be it in France or India), levels of lactase remain adequate through life (Simoons 1979; Kozlov et al. 1998).

But even populations with high prevalence of lactase deficiency can change their diet and become substantial consumers of dairy products, something they would not obviously do if it were accompanied by quotidian experience of abdominal discomfort, or even vomiting. Three realities explain this: lactase synthesis is completely turned off in only a small minority of adults while others can enjoy milk as long as they do not drink it at Dutch or Danish rates; surprisingly, bacterial enzymes present in fermented dairy products (including buttermilk, sour cream and yogurt) make their digestion easier by providing substitutes for the endogenous lactase; and, depending on their level of ripeness, cheeses have either much lower levels of lactose than milk (unripe varieties) or (all fully ripened hard varieties) only its merest trace (Hui 1993). Consumption of dairy products is thus more limited by tradition, taste or expense, and Japan, a country where essentially no milk was drunk and no yogurt or cheeses were sold until after WW II, is certainly the best example of how fast and how far even those three obstacles can change.

By 1960, the national average of milk drinking had quintupled to 20 L/capita, mainly due to the National School Lunch Program, and by 1970 it more than doubled to 45 L/capita. Not surprisingly, Takahashi (1984) saw this increased milk consumption as a major reason for an unprecedented growth in average height of children between 1960 and 1975. The average intake had eventually plateaued at around 65 L/capita; by the year 2000, Japan's annual supply of all dairy products was equal to 95 L of milk, and a survey in 2002 found that only 11% of all respondents did not drink any milk and 15% did not eat any yogurt (Watanabe and Suzuki 2006).

America's consumption of dairy products (in terms of milk equivalent) peaked during the late 1930s and 1940s and has been declining ever since, decreasing from 385 L/capita to about 275 L/capita by 2010 as falling intakes of milk were only partially compensated by higher demand for cheeses, from about 2 kg/year in 1945 to about 15 kg/year in 2010 (USDA 2012a). In 2010, the average per capita supply of protein from dairy products was about 20 g/day compared to about 45 g/day for red and white meat. Returning the average US per capita consumption of dairy products to its record level (i.e., raising it by 110 L/year or by about 40%) would add about 8 g of animal protein a day and that would displace about 18% of meat protein supply.

Consequently, it would be realistic to envisage that the US can displace 15–20% of its meat protein by dairy protein, and dividing the increased dairy consumption among whole and low-fat milk, yogurt, creams, cheeses and iced products would not call for any drastic shifts in the prevailing diet: it would be equivalent to eating a standard portion of each of these foods four or five days a week rather than just three days a week. Increasing consumption of milk in China, traditionally a non-milking society, has already made a contribution and can go much further in order to reduce the future demand for meat. Half a century ago, China's per capita milk supply was minuscule, at less than 1 L/year, and by 2010, mainly due to increasing urban demand, it was nearly 30 L/year, and hence its consumption could roughly triple before it would reach the recent Japanese rate: at that level, dairy products would provide nearly 10 g of protein a day, compared to almost 20 g of protein a day now originating from meat.

As in the case of milk, increased consumption of eggs would make a greater difference to protein supply in countries with low to moderate meat intakes. In the US, eggs now provide less than 10% of all protein supplied by red and white meat, and hence replacing just 10% of all meat protein by egg whites would require the doubling of average annual egg consumption to about 500 eggs (currently about 170 eggs are eaten annually in shell, 80 in products). That would be a level never reached in any country: in the US, annual per capita egg consumption peaked during the late 1940s at more than 400 eggs, and recently the world's top per capita consumers (Mexico, China and Japan) were all below 350 eggs/year (FAO 2012). Consequently, a share of the existing meat protein intake that could be realistically displaced by eggs is no more than 5% in nations with high meat consumption but easily 10% in many Asian and African countries.

This leaves us to consider meat substitutions by aquatic species, perhaps the most desirable of the three possible shifts because standard dietary recommendations abound with admonitions to consume less red meat and to eat more fish. But unlike in the case of dairy products and eggs – where a fraction of feed used to produce meat could be easily diverted to boost their output – it is not only impossible to increase ocean catches by 10% or

20%, but in the short- and medium-term, there should be actually substantial cuts to aid the recovery of overfished stocks. The parlous state of global fisheries has been well documented as virtually all principal fishing regions and all commercially valuable species are either overexploited – with some species, including Canada's Atlantic cod, still not recovering from collapse of stocks – or fished to capacity (FAO 2010d). As a result, the global ocean catch has been stagnating around 80 Mt/year (2005: 82.8 Mt, 2009: 78.6 Mt), and it is most unlikely that it will resume its pre-1990 upward trend.

Moreover, any intensification of fishing would also worsen the overall environmental impact by causing more harm to species not targeted for capture: unfortunately, fishing does not reduce only the stocks of targeted species, but a great deal of collateral damage is done because of by-catch that includes unwanted fish, invertebrates, reptiles (turtles ensnared in fishing nets), mammals (perhaps most common victims are dolphins caught in tuna nets) and even seabirds (albatrosses) killed by long-lines set for tuna. This by-catch is discarded over board, as are targeted species that are undersized, females that are protected for breeding as well as any catch that exceeds allotted quota.

The extent of by-catch varies widely, in terms of mass from as little as 10% in octopus fishery to more than 500% in some shrimping, with rates between 60% and 75% common for pelagic and demersal fish, and the global mean during the early 1990s was at least 35%, or nearly 30 Mt/year in 1990 (Alverson et al. 2004). Subsequent improvements in fishing operations have lowered the by-catch to about 15 Mt/year (Alverson 2005) or to as low as 7.3 (6.8–8.0) Mt/year (Kelleher 2004). But discard mortality remains high, and its rates are highly species-specific, ranging from less than 5% to more than 80% for such valuable fish as halibut and salmon (Alverson et al. 2004).

The latest assessment by the FAO concluded that in order to return fishing to a sustainable level, by the year 2015 the worldwide fishing effort would have to be cut back by 36–43% (Ye et al. 2012). Landings would be cut by several tens of million tons a year, 12–15 million fishers would lose work and more than $300 billion would be taken out of the world economy, losses that would not be perceived by the affected parties as bearable just because the catches might (or may not) increase in the long run. In any case, substantial near-term gains in fish supply will have to come from freshwater and marine aquaculture that can produce high-quality protein with superior feed conversion efficiencies.

Fish convert feed with much higher efficiency than mammals or birds because they are ectothermic and do not waste any energy on maintaining steady body temperature, because their buoyant surroundings and stream-lined body shape reduce the cost of locomotion (note how small those myoglobin-infused muscles are), and because they do not need energy to convert ammonia to urea and uric acid (their main excretory pathway is ammonia through the gills). Feed conversion efficiencies are thus rela-tively high for all aquacultured fish, but there is an important difference between herbivorous and carnivorous species as far as the composition of feed is concerned.

Herbivorous species raised in warm climates can be fed concentrate mixtures very similar to those fed to pigs or broilers, containing pellets made of corn (or other grains) and soybeans (and other oilseeds), and they will need as little as 1.1 units of feed per unit of gain; even with lower-quality feeds, the ratio is no higher than 1.6. Omnivorous fish (grass carp, common carp, catfish, tilapia, milkfish) have higher protein requirements (less than 20%), and about 5% of this is supplied as fishmeal. That is why freshwater herbivores and omnivores are the leading categories of modern aquaculture, and why in China, the world's leading aquacultural producer (accounting for nearly two-thirds of the global output that surpassed 55 Mt in 2009), four species of carp (herbivorous silver and bighead and omnivorous grass and common carps) are by far the most important farmed fish.

Those four species are also among the top six products of global aqua-culture by mass (Japanese carpet shell and Nile tilapia taking the third and fifth place in 2009), and their annual production adds up to a quarter of the worldwide aquacultural output (FAO 2012). Unfortunately, most societies do not share China's fondness of carp – and nor are they keen on eating herbivorous herring whose relatively high consumption remains restricted to colder Atlantic Europe (even Japanese are not that keen on *nishin*). Hence, the food species that dominate commercial ocean catches are either planktivores (tertiary consumers preying on zooplankton) or piscivores (preying on fish).

Pollock (eaten overwhelmingly only after processing *surimi*, imitation crab meat), pink and sockeye salmon belong to the first category, while cod, Chinook and coho salmon and different tunas are the most favored piscivorous fish choices. For example, America's nationwide dietary intake study found that only about 2% of all adults reported eating sardines (the cheapest of all inexpensive herbivorous fish) and 1% eating mackerel (a small piscivore) during the 30 days preceding the inquiry, with carnivo-rous tuna and salmon topping the frequency list with, respectively, 35% and 27% (CDCP 2011). Not surprisingly, stocks of those much-demanded species are in a particularly bad shape.

Cod has been both a European and North American favorite for centuries, but its largest Atlantic fishery off Newfoundland and Nova Scotia (also one of the world's oldest continuous endeavors of its kind with European fleets coming to those waters since the beginning of the 16th century) had collapsed in 1985. North Sea cod numbers have declined greatly; most salmon runs have been either severely depleted or amount to small shares of their intensity of a few decades ago, and the most sought-after species of tuna (particularly the Pacific bluefin thanks to Japan's high demand and willingness to pay exorbitant prices for choice *maguro*) have been under growing pressure around the world, and particularly in the Mediterranean and the Eastern Atlantic.

These realities have turned marine aquaculture into a major commercial activity whose annual output has been steadily increasing, and with an output of just over 8 Mt in 2009, it was equal to about 12% of all ocean catches (FAO 2010d). Leading products in terms of zoomass and value are, respectively, molluscs (mainly the Pacific oyster, *Crassostrea gigas*) and crustaceans (shrimps and prawns), while salmonids (salmon and trouts) top the marine fish list. Even bluefin tuna is now grown in captivity as young fish are caught in wild and placed in feeding cages. Carnivorous species need 20–40% of their diet either as fishmeal or as a combination of fishmeal and fish oil, and during the 1970s early aquaculture of Atlantic salmon had feed conversion ratios higher than 2, but by the late 1990s the typical performance was around 1.5. Today, the average is just 1.3, with the best operations just around 1.0.

Recent average performance is double the rate in large broiler operations and more than three times the rate for pork – but while birds and mammals convert plants to meat, a substantial share of feed for piscivorous fish must come from fishmeal (currently at least a quarter of the total), and for salmon there must also be adequate supply of fish oil in feed, typically between 15% and 20% (Naylor et al. 2009). Fishmeal needs are similar (18–20%) for crustaceans or even higher (30%) for tuna whose mariculture is (in trophic-level terms) a terrestrial equivalent of raising tigers for meat (Naylor and Burke 2005). As a result, recent fishmeal consumption in aquaculture has approached 4 Mt/year, or nearly two-thirds of the bulk of global fishmeal output, with salmonids and crustaceans each consuming about a quarter of the total (FAO 2010d).

Fortunately – and contrary to a common perception – this feeding does not reduce the overall supply of edible fish. Several species used to produce fishmeal have no market as food (Gulf and Atlantic menhaden, sand eel), others claim only very small localized food market niches (anchovies, capelin, sprats) while some catches of food fish (notably some mackerel and sardine species) are rejected by the market, and if these would not be converted to feed, they would be wasted. Nor is it true that leaving the fish for wild species to feed on would produce more protein: feed conversion efficiencies of species at higher trophic level feeding on the species used for fishmeal are significantly lower than in salmonid or crustacean aquaculture (Wijkström 2009).

Moreover, plant substitutes (including oils derived from soybeans and rapeseed) can replace part of fish oil, and breeding has already lowered the demand for fishmeals and oils in fish feed: global aquaculture production has continued to grow, while the demand for fish-derived feeds has remained flat. But the need for feeding fish to produce piscivorous fish favored by consumers will never be eliminated, and, hence, even when assuming that it will be no more than 15% of total feed mass, the fishmeal output will double thanks to a near-complete use of fish processing wastes (currently only about a quarter of those wastes is rendered).

In any case, there are obvious limits to marine aquaculture as a supplier of high-quality protein that could displace meat consumption even when assuming that most consumers would be willing to make the switch (with price, not just taste, to be considered). Global mariculture still amounts to less than 10 Mt/year (compared to less than 70 Mt/year for marine capture), and even a tripling of that total during the next generation (perhaps as much as could be realistically envisioned) would produce annually net protein equivalent (after discards and processing for retail sales) to no more than 25 Mt of meat, displacing less than 9% of 2010 global meat production, and (using FAO's forecast of about 370 Mt) less than 7% of global demand in 2030, but a greatly expanded freshwater aquaculture could bring that share closer to 10%. How much more marine protein could be eventually caught by recovered and properly managed fisheries is too uncertain to estimate.

Adding up these realistic substitutions shows that diverting feeds into more efficient ways of producing animal protein could displace at least 25% of today's meat supply in European and North American affluent countries with high levels of meat consumption (15% by dairy products, 5% each by eggs and seafood) and up to 40% in Asian countries with moderate meat intake and traditional preference for freshwater fish (20% by dairy products and 10% each by eggs and seafood). Even the lowest national substitution rates should be at least 15%, and the highest realistic displacement could be as much as 50% of today's average per capita meat intakes. Of course, these shares are calculated by using the current levels of meat consumption: obviously, they would be lower if the demand for meat continues to rise – but higher if it followed a declining trend as many societies undergo yet another dietary transition to less meaty diets.

Less meaty diets

After the three post-WW II generations of virtually continuous growth of all production and consumption indicators, standard expectations have been for more of the same during the coming decades. The only difference is that slower growth rates have been generally envisaged for affluent countries with slowly growing or stagnating (and in the future even declining) populations, with the global growth (be it of energy, food or manufactured products) dominated by the demand in modernizing countries of Asia and Latin America (eventually to be joined by most of Africa). Not surprisingly, forecasts of global meat consumption follow the same pattern: FAO (2005) estimated the global total at 373 Mt in 2030 (compared to 234 Mt in the year 2000), with only a third of it in today's affluent economies.

Even a modest rise in disposable incomes in the world's major modernizing economies with currently low or very low meat intakes would translate into higher future demand, and only a preservation of today's, still widespread, level of poverty and income inequality would prevent that development. But it is not at all clear that meat demand will be, however modestly, rising in high-income nations where the post-1990 statistics indicate that the average per capita supply of meat in general, and of red meats in particular, has either already reached a plateau or is very close to it (at about 120 kg/year in the US, 100 kg in Australia and between 80 and 90 kg in the EU), and that in some cases a new trend of modest declines is under way.

Japan's post-WW II per capita meat supply (carcass weight) rose rapidly by an order of magnitude, from less than 5 kg during the late 1950s to about 45 kg by the year 2000; in edible terms, the latter total was about 29 kg of meat, but during the next decade the average per capita supply had stabilized at about 28 kg/year (Smil and Kobayashi 2012). Given the country's aging, and more health-conscious, population, it is almost certain that average per capita meat consumption has already peaked and that its further slow decline is ahead. Reduced eating of meat would be no particular hardship for most Europeans: they have been moving in that direction for a few decades.

Few people outside France realize that even most of the meat-loving French have become *petits consommateurs*. In 2007, a survey showed that 56% of French eat less than 45 g of meat a day, that is, no more than about 16.5 kg a year, and only 20% eat more than 70 g a day or 25 kg a year (CIV 2009). And, surprisingly, France has also seen the fastest recent decline in meat consumption among the richest populous nations: dietary surveys of adults (15+ years) show their daily average consumption of all meat and meat products (*charcuterie, produits carnés des plats préparés*) falling from 147 g in 1999 to 122 g in 2004 and 117 g in 2007, a 20% decline in just eight years (CIV 2009).

For red meats (*viandes de boucherie*), the decline was slightly higher from 59 to 46 g/day, and a slightly different survey for adults over 18 years of age showed a 15% decline in red meat consumption in the seven years between 2003 and 2010 (Hébel 2012). The steepness of this decline makes it unclear for how long this trend will continue: if uninterrupted, French meat consumption in 2025 would be nearly 30% lower than in the year 2000. After recovering from post-WW II destruction, German meat supply (carcass weight) rose from just over 60 kg in 1960 to 96 kg by 1980 and then peaked at around 100 kg during the late 1980s before declining to less than 90 kg by the year 2010: this means that it took a generation to lower the demand by some 12%, and, as in the case of Japan, further slow decline is the most likely expectation.

And the US per capita meat supply, relatively very high even in 1900, grew by 50% between the first and the last decade of the 20th century, but then it declined by less than 4% between the years 2000 and 2009 (USDA 2012a). Since 1970, per capita pork consumption has fluctuated within a narrow range just above 20 kg/year; beef consumption peaked in 1975–1976 (at just over 40 kg of retail weight), fell to less than 31 kg by 1990, leveled off for a while, and its decline was further accentuated by the post-2007 economic downturn: in 2011, American per capita beef consumption was just 26 kg, 13% less than in 2001 and 25% less than in 1980, and the USDA expected further slow decline (Wells and Buzby 2008). As a result, the mean US per capita demand for all meat has reached a clear plateau of around 120 kg/capita, and it is quite possible that it will not surpass this level. Only chicken consumption has been rising: it surpassed beef intake in 1992, and two years later the consumption of chicken away from home surpassed the household servings.

Substantially higher meat prices (or lower disposable incomes) would certainly accelerate declines in meat demand that have been, so far, driven mostly by changing food preferences due to health concerns and population aging. Studies of price elasticity of meat – be they for specific meat varieties or for an entire national meat demand – can be used to estimate the magnitude of price increases that would translate into substantial declines in meat demand, and Gallet (2009) has made the task easier by collecting 419 publications of that kind for his meta-analysis. Median price elasticity for a total of 4,142 estimates for all meat varieties was −0.774, but, as expected, the standard deviation of 1.28 indicates a great deal of variation. In terms of regional elasticities, East Asia's value was virtually identical to the overall mean (–0.77), and the range was from –1.054 for Western Europe to –0.463 for South America.

As for the individual varieties, nearly 1,000 observations for beef resulted in mean price elasticity of –0.869, while the elasticities for pork and poultry were, respectively, –0.780 and –0.650. For comparison, estimated long-term compensated elasticities for meat demand in the US were –0.61 for beef, –0.5 for pork and –0.05 for poultry (Schroeder et al. 2000). A policy that would single out beef – because of its relatively low feed conversion ratio or relatively high overall environmental impact – for a substantial price-driven reduction in demand would thus require, with the most likely price elasticity around –0.75, that the price would have to be one-third higher in order to reduce the demand for that meat by about 25%. Overall cut in meat consumption would be slightly lower as positive cross-price elasticities for beef and pork indicate that they are partial substitutes: 1% increase in US beef price would result in almost 0.1% increase in pork demand and 0.05% increase in poultry consumption.

Eventual saturation levels of the Asian, and later also of the African, meat demand are much more difficult to predict. Interestingly (and largely thanks to small declines of urban meat demand), the growth of China's overall supply has already slowed down less than three decades after the onset of the country's rapid post-1980 dietary transition, with the per capita average rising by only about 10% between 2000 and 2009. This relatively modest gain is less surprising once it is realized that the Chinese per capita meat (i.e., mostly pork) consumption rate is already close to that of South Korea and not far behind Taiwan, and that long-term income elasticity of China's pork consumption is only about 0.15 (Masuda and Goldsmith 2010).

Liu et al. (2009) documented a twofold effect of rising incomes in China: increase in the total consumption and changed composition of intakes away from traditional pork by including more poultry and beef. And because long-term income elasticities for poultry and beef consumption are, respectively, 1.057 and 1.560, Masuda and Goldsmith (2010) estimate that, compared to 2010, in 2030 China will consume more than twice as much poultry and three times as much beef and that the country's total per capita meat consumption (carcass weight) will surpass 90 kg/year, or more than is now the EU average. While that forecast may turn out to be a major exaggeration, even increases equal to half the projected rate for poultry and beef would amount to additional 20 Mt of carcass weight, roughly an equivalent of Brazil's total meat production in 2010.

And after tripling between 1960 and 2010, Brazil's meat supply (nearly half of it beef) is above 80 kg/capita (FAO 2012). Although that is essentially at the EU level, it might be premature to think that a saturation level has been reached. Given the country's notorious income inequalities and lasting disparities between the quality of urban and rural food consumption and among the different socioeconomic classes, further increases might be possible once the disposable incomes of the poorest groups increase. After all, as Oths et al. (2005, 319) put it, "the poor are shut out from participation in one of the most fundamental activities signifying their national identity – eating meat. It is particularly ironic … that the class that gave the country *feijoada* today cannot afford to eat it."

And the future of India's meat consumption is particularly hard to foresee. How many lacto-vegetarians, now the majority of the country's population, will become at least occasional consumers of poultry or goat meat? How much more beef will India's large Muslim population consume? Long-term comparisons show that meat demand slows considerably once the average annual per capita income goes over $7,000 and that it more or less levels off after $10,000, but most of the world's populous modernizing nations are still far below even the first threshold. As a result, the saturation of average per capita meat demand followed by a gradual decline in affluent and middle-income countries will not result in an overall reduction of global meat demand during the next generation. Even if by 2030 average capita supply for 1.3 billion affluent people were to decline by as much as 25%, such a decline would be far too small to make up for only a 10% higher demand for 7 billion people that are expected to live at that time in Asia, Africa and Latin America.

And so (barring an unpredictable, severe and prolonged worldwide economic downturn or an unprecedented deterioration of global environment), even reduced meat demand in high-income societies combined with a relatively slow rise of meat consumption in modernizing countries will require expanded meat production. But, as I have just demonstrated by approximate quantifications of realistic meat substitution, the future totals of global demand for meat could be significantly lower than those forecast by the FAO. As the final step of this exploration of alternative futures, I will calculate how much meat we could produce with minimal impacts – and compare with both the most likely and the most rational rates of demand.

A large potential for rational meat production

The most obvious path toward more rational meat production is to improve efficiencies of many of its constituent processes and hence reduce waste and minimize many undesirable environmental impacts. As any large-scale human endeavor, meat production is accompanied by a great deal of waste and inefficiency, and while he have come close to optimizing some aspects of the modern meat industry, we have a long way to go before making the entire enterprise more acceptable. And, unlike in other forms of food production, there is an added imperative: because meat production involves breeding, confinement, feeding, transportation and killing of highly evolved living organisms able to experience pain and fear, it is also accompanied by a great deal of unnecessary suffering that should be eliminated as much as possible.

Opportunities to do better on all of these counts abound, and some are neither costly nor complicated: excellent examples range from preventing the stocking densities of pastured animals from surpassing grassland's long-term carrying capacity to better designs for moving cattle around slaughterhouses without fear and panic. There is no shortage of prescriptions to increase global agricultural production with the maintenance of well-functioning biosphere or, as many of my colleagues would say, to develop sustainable food production while freezing agriculture's environmental footprint of food (Clay 2011) – or even shrinking it dramatically (Foley et al. 2011). As always, I am reluctant to use the adjective "sustainable" (the term remains poorly defined, and it is commonly used to describe actions that, at a closer look, are not at all truly sustainable). Nor do I believe, given the inherent functional interconnectedness governing the biosphere, in defining assorted footprints (carbon, water, nitrogen) for specific human activities.

In any case, the two key components in the category of improvements are the effort to close yield gaps due to poor management rather than to inferior environmental limitations and to maximize the efficiency with which the key resources are used in agricultural production. Claims regarding the closing of the yield gaps must be handled very carefully as there are simply too many technical, managerial, social and political obstacles in the way of replicating Iowa corn yield throughout Asia, to say nothing about most of sub-Saharan Africa, during the coming generations. Africa's average corn yield rose by 40% between 1985 and 2010 to 2.1 t/ha, far behind the European mean of 6.1 and the US average of 9.6 t/ha, but even if it were double during the next 25 years to 4.2 t/ha, the continent's continuing rapid growth would reduce it to no more than about 35% gain in per capita terms. Asian prospects for boosting the yields are better, but in many densely populated parts of that continent, such yields might be greatly reduced, even negated by the loss of arable land to continuing rapid urbanization and industrialization.

At the same time, there does not appear to be anything in the foreseeable future that could fundamentally change today's practices of growing livestock for meat. Indeed, many arguments can be made that after half a century of focused breeding, accelerated maturation of animals and improvements in feed conversion, these advances have gone too far and are now detrimental to the well-being of animals and to the quality of the food chain and have raised environmental burdens of meat production to an unprecedented level that should not be tolerated in the future. And neither the expanded aquaculture nor plant-based meat imitations will claim large shares of the global market anytime soon, and cultured meat will remain (for a variety of reasons) an oddity for a long time to come.

Consequently, it is very unlikely (again, barring an unprecedented global economic depression) that the undoubted, continuing (and possibly even slightly accelerating) positive impact of the combination of higher productivities, reduced waste, better management and alternative protein supplies would make up for additional negative impacts engendered by rising meat production and that there would be discernible net worldwide improvement: the circle of reduced environmental impacts cannot be squared solely by more efficient production. At the same time, the notion that an ideal form of food production operating with a minimal environmental impact should exclude meat – nothing less than enacting "vegetarian imperative" (Saxena 2011) on a global scale – does not make sense.

This is because both grasslands and croplands produce plenty of phytomass that is not digestible by humans and that would be, if not regularly harvested, simply wasted and left to decay. In addition, processing of crops to produce milled grains, plant oils and other widely consumed foodstuffs generates a large volume of by-products that make (as described in Chapter 4) perfect animal feeds. Rice milling strips typically 30% of the grain's outermost layers, wheat milling takes away about 15%: what would we do with about 300 Mt of these grain milling residues, with roughly the same mass of protein-rich oil cakes left after extraction of oil (in most species accounts for only 20–25% of oilseed phytomass), and also with the by-products of ethanol (distillers grain) and dairy industries (whey), waste from fruit and vegetable canning (leaves, peels), and citrus rinds and pulp?

They would have to be incinerated, composted or simply left to rot if they were not converted to meat (or milk, eggs and aquacultured seafood). Not tapping these resources is also costly, particularly in the case of porcine omnivory that has been used for millennia as an efficient and rewarding way of organic garbage disposal. Unfortunately, in 2001, the EU regulations banned the use of pig swill for feeding, and Stuart (2009) estimated that this resulted in an economic loss of €15 billion a year even when not counting the costs of alternative food waste disposal from processors, restaurants and institutions. Moreover, the ban has increased CO_2 emissions as the swill must be replaced by cultivated feed.

At the same time, given the widespread environmental degradation caused by overgrazing, the pasture-based production should be curtailed in order to avoid further soil and plant cover degradation. Similarly, not all crop residues that could be digested by animals can be removed from fields, and some of those that can be have other competing uses or do not make excellent feed choices, and not all food processing residues can be converted to meat. This means that a realistic quantification of meat production potential based on phytomass that does not require any cultivation of feed crops on arable land cannot be done without assumptions regarding their final uses, and it also requires choices of average feed conversion ratios. As a result, all such calculations could be only rough approximations of likely global totals, and all of my assumptions (clearly spelled out) err on a conservative side.

Because most of the world's grasslands are already degraded, I will assume that the pasture-based meat production in low-income countries of Asia, Africa and Latin America should be reduced by as much as 25%, that there will be absolutely no further conversion of forests to grasslands throughout Latin America or in parts of Africa, and that (in order to minimize pasture degradation in arid regions and nitrogen losses from improved pastures in humid areas) grazing in affluent countries should be reduced by at least 10%. These measures would lower pasture-based global beef output to about 30 Mt/year and mutton and goat meat production to about 5 Mt.

Another way to calculate a minimum production derived from grasslands is to assume that as much as 25% of the total area (the most overgrazed pastures) should be taken out of production and that the remaining 2.5 Gha would support only an equivalent of about half a livestock unit (LU, or roughly 250 kg of cattle live weight) per hectare (for comparison, since 1998 the EU limits the grazing densities to 2 LU/ha, Brazil's grasslands typically support 1 LU/ha and 0.5 LU is common in sub-Saharan Africa). Assuming average annual 10% off-take rate and 0.6 conversion rate from live to carcass weight, global meat production from grazing would be close to 40 Mt/year, an excellent confirmation of the previous total derived by different means.

At the same time, all efforts should be made to feed available crop residues to the greatest extent possible. Where yields are low and where the cultivated land is prone to erosion, crop residues should be recycled in order to limit soil losses, retain soil moisture and enrich soil organic matter. But even with much reduced harvest ratios of modern cultivars (typically a unit of straw per unit of grain), high yields result in annual production of 4–8 t of straw or corn stover per hectare, and a very large part of that phytomass could be safely removed from fields and used as ruminant feed. The annual production of crop residues (dominated by cereal straws) now amounts to roughly 3 Gt of dry phytomass.

Depending on crops, soils and climate, recycling should return 30–60% of all residues to soil, and not all of the remaining phytomass is available for feeding: crop residues are also used for animal bedding; for many poor rural families in low-income countries, they are the only inexpensive household fuel; and in many regions (in both rich and poor countries) farmers still prefer to burn cereal straw in the fields – this recycles mineral nutrients but it also generates air pollution. Moreover, while oat and barley straws and stalks and leaves of leguminous crops are fairly, or highly, palatable, ruminants should not be fed solely by wheat or rice straw; rice straw in particular is very high in silica (often in excess of 10%), and its overall mineral content may be as high as 17%, more than twice that of alfalfa. As a result, the best use of cereal straws in feeding is to replace a large share (30–60%) of high-quality forages.

These forages should be cultivated preferably as leguminous cover crops (alfalfa, clovers, vetch) in order to enhance the soil's reserves of organic matter and nitrogen. If only 10% of the world's arable land (or about 130 Mha) were planted annually with these forage crops (rotated with cereals and tubers), then even with a low yield of no more than 3 t/ha of dry phytomass, there would be some 420 Mt of phytomass available for feeding, either as fresh cuttings or as silage or hay. Matching this phytomass with crop residues would be quite realistic as 420 Mt would be only about 15% of the global residual phytomass produced in 2010. Feeding 840 Mt of combined forage and residue phytomass would, even with a very conservative ratio of 20 kg of dry matter/kg of meat (carcass weight), produce at least 40 Mt of ruminant meat.

Unlike in the case of crop residues, most of the food processing residues are already used for feeding, and the following approximations quantify meat production based on their conversion. Grain milling residues (dominated by rice and wheat) added up to at least 270 Mt in 2010, and extraction of oil yielded about 310 Mt of oil cakes. However, most of the latter total was soybean cake whose output was so large because the crop is now grown in such quantity (about 260 Mt in 2010) primarily not to produce food (be it as whole grains, fermented products including soy sauce and bean curd, and cooking oil) but as a protein-rich feed.

When assuming that soybean output would match the production of the most popular oilseed grown for food (rapeseed, at about 60 Mt/year), the worldwide output of oil cakes would be about 160 Mt/year. After adding less important processing by-products (from sugar and tuber, and from vegetable and fruit canning and freezing industries), the total dry mass of highly nutritious residues would be about 450 Mt/year of which some 400 Mt would be available as animal feed. When splitting this mass between broilers and pigs, and when assuming feed : live weight conversion ratios at, respectively, 2 : 1 and 3 : 1 and carcass weights of 70% and 60% of live weight, feeding of all crop processing residues would yield about 70 Mt of chicken meat and 40 Mt of pork.

The grand total of meat production that would come from grazing practiced with greatly reduced pasture degradation (roughly 40 Mt of beef and small ruminant meat), from feeding forages and crop residues (40 Mt of ruminant meat) and from converting highly nutritious crop processing residues (70 Mt chicken meat and 40 Mt pork) would thus amount to about 190 Mt/year. This output would require no further conversions of forests to pastures, no arable land for growing feed crops, no additional applications of fertilizers and pesticides with all the ensuing environmental problems. And it would be equal to almost exactly two-thirds of some 290 Mt of meat produced in 2010 – but that production causes extensive overgrazing and pasture degradation, and it requires feeding of about 750 Mt of grain and almost 200 Mt of other feed crops cultivated on arable land predicated on large inputs of agrochemicals and energy.

And the gap between what I call rational production and the actual 2010 meat output could be narrowed. As I have used very conservative assumptions, every component of my broad estimate could be easily increased by 5% or even 10%. Specifically, this could be achieved by a combination of slightly higher planting of leguminous forages rotated with cereals, by treatment of straws with ammonia to increase its nutrition and palatability, by a slightly more efficient use of food processing by-products and also by elimination of some of the existing post-production meat waste. Consequently, the total of 200 Mt/year can be taken as an unassailably realistic total of global meat output that could be achieved without any further conversion of natural ecosystems to grazing land, with conservative pasture management, and without any direct feeding of grains (corn, sorghum, barley), tubers or vegetables, that is, without any direct competition with food produced on arable land.

This amounts to almost 70% of the actual meat output of about 290 Mt in the year 2010: it would not be difficult to adjust the existing system in the described ways, eliminate all cultivation of feed crops on arable land (save for the beneficial rotation with leguminous forages) and still average eating only a third less meat than we eat today. And, as you will recall from the calculations of meat displacements by increased production of other animal foodstuffs, they could realistically supply an equivalent of 15–20% of the current meat output (with much higher feed conversion efficiency and with much less environmental impact). This means that the combination of rational meat production and realistically projected meat displacements by other animal foodstuffs would produce 85–90% of high-quality protein available in 2010 global meat output.

A key question to ask then is how the annual total of some 200 Mt of meat would compare with what I would term a rational consumption of meat rather than with the existing level. Making assumptions about rational levels of average per capita meat consumption is done best by considering actual meat intakes and their consequences. As already noted, a slight majority of people in France, the country considered to be a paragon of classic meat-based cuisine, now eat no more than about 16 kg of meat a year per capita, and the average in Japan, the nation with the longest life expectancy, is now about 28 kg of meat (both rates are for edible weight). Consequently, I will round these two rates and take the per capita values of 15–30 kg/year as the range of rational meat consumption. For seven billion people in 2012, this would translate to between 105 and 210 Mt/year – or, assuming 20/30/50 beef/pork/chicken shares, between 140 and 280 Mt in carcass weight. The latter total is almost equal to the actual global meat output in 2010, with the obvious difference being that the consumption of today's output is very unevenly distributed.

If we could produce 200 Mt/year without any competition with food crops, then the next step is to inquire how much concentrate feed we would need to grow if we were to equal current output of roughly 300 Mt with the lowest possible environmental impact. Assuming that the additional 100 Mt meat a year would come from a combination of 10 Mt of beef fed from expanded cultivation of leguminous forages, 10 Mt of herbivorous fish (conversion ratio 1 : 1) and 80 Mt of chicken meat (conversion ratio 2 : 1), its output would require about 170 Mt of concentrate feed, that is, less than a fifth of all feed now produced on arable land. Moreover, a significant share of this feed could come from extensive (low-yield and hence low-impact) cultivation of corn and soybeans on currently idle farmland.

Roques et al. (2011) estimated that in 2007 there were 19–48 Mha of idle land (an equivalent of 1.3–3.3% of the world's arable area), that is, land cultivated previously that can be planted again, most of it in North America and Asia. Using 20 Mha of this land would produce at least an additional 60 Mt of feed. And when factoring in increasing crop yields, regular rotations with leguminous forages (producing excellent ruminant feed while reducing inputs of nitrogen fertilizers) and, eventually, slightly higher feed conversion efficiencies, it is realistic to expect that the share of the existing farmland used to grow feed crops could be reduced from the current share of about 33% to less than 10% of the total. Consequently, there is no doubt that we could match recent global meat output of about 300 Mt meat a year without overgrazing, with realistically estimated feeding of residues and by-products, and with only a small claim on arable land, a combination that would greatly limit livestock's environmental impact.

Prospects for Change

What are we to make of these comparisons? I must stress that I did not introduce them to be just intriguing but essentially irrelevant exercises based on theoretical calculations that ignore the realities of actual meat demand and meat production. They are solid, conservative quantifications of possible alternatives that show that other options – some relatively easily achievable, others requiring longer transition – are available and could be pursued, and that those pursuits would be rewarded by numerous environmental, economic and health benefits. But will we pursue them, and if so, with what determination and how persistently?

Many years ago, I decided not to speculate about the course and intensity of any truly long-term developments: all that is needed to show a near-complete futility of these efforts is to look back and see to what extent would have any forecast made in 1985 captured the realities of 2010 – and that would be looking just a single generation ahead, while forecasts looking half a century into the future are now quite common. Forecasting demand for meat – a commodity whose production depends on so many environmental, technical and economic variables and whose future level of consumption will be, as in the past, determined by a complex interaction of population and economic growth, disposable income, cultural preferences, social norms and health concerns – thus amounts to a guessing game with a fairly wide range of outcomes.

But FAO's latest long-range forecast gives just single global values (accurate to 1 Mt) not just for 2030 (374 Mt) but also for 2050 (455 Mt) and 2080 (524 Mt). Compared to 2010, the demand in 2030 would be nearly 30%, and in 2050 about 55% higher. When subdivided between developing and developed countries, the forecast has the latter group producing in 2080 only a third as much as the former. These estimates imply slow but continuing growth of average per capita meat consumption in affluent countries (more than 20% higher in 2080 than in 2007) and 70% higher per capita meat supply in the rest of the world.

Standard assumptions driving these kinds of forecasts are obvious: either a slow growth or stagnation and decline of affluent population accompanied by a slow increase of average incomes; continuing, albeit slowing, population growth in modernizing countries where progressing urbanization will create not only many new large cities but also megacities, conurbations with more than 20 or 30 million people, and boost average disposable incomes of billions of people; advancing technical improvements that will keep in check the relative cost of essential agricultural inputs (fertilizers, other agrochemicals, field machinery) and that will keep reducing environmental impacts; and all of this powered by a continuing supply of readily available fuels and electricity whose cost per unit of final demand will not depart dramatically from the long-term trend.

Standard assumptions also imply continuation and intensification of existing practices ranging from large-scale cultivation of feed crops on arable land (with all associated environmental burdens) to further worldwide diffusion of massive centralized animal feeding operations for pork and poultry. Undoubtedly, more measures will be taken to improve the lot of mammals and birds in CAFOs. Many of them will be given a bit more space, their feed will not contain some questionable ingredients, an increasing share of them will be dosed less with unnecessary antibiotics and their wastes will be better treated. Some of these changes will be driven by animal welfare considerations, others by public health concerns, new environmental regulations and basic economic realities; all of them will be incremental and uneven. And while they might be cumulatively important, it is unlikely that their aggregate positive impact will be greater than the additional negative impact created by substantial increases in the expected demand for meat: by 2030 or 2050, our carnivory could thus well exact an even higher environmental price than today.

I have tried to demonstrate that such an outcome is not foreordained and that alternative courses are not just imaginable but quite possible. To begin with, I would strongly argue that there is absolutely no need for higher meat supply in any affluent economy, and I do not think that improved nutrition, better health and increased longevity in the rest of the world is predicated on nearly doubling meat supply in today's developing countries. I have shown that global output of as little as 140 Mt/year (carcass weight) would guarantee minimum intakes compatible with good health, and production on the order of 200 Mt of meat a year could be achieved without claiming any additional grazing or arable land and with water and nutrient inputs no higher than those currently used for growing just food crops.

And it could also be done in a manner that would actually improve soil quality and diversify farming income. Moreover, an additional 100 Mt/year could be produced by using less than a fifth of the existing harvest of concentrate feeds, and it could come from less than a tenth of the farmland that is now under cultivation and that could be used to grow food crops. Even for a global population of eight billion, the output of 300 Mt/year would prorate to nearly 40 kg of meat a year/capita, or well above 50 kg a year for adults. This means that the average for the most frequent meat eaters, adolescent and adult men, could be 55 kg/year, and the mean for women, children and people over 60 would be between 25 and 30 kg/year, rates that are far above the minima needed for adequate nutrition and even above the optima correlated with desirable health indicators (low obesity rates, low CVD mortality) and with record nationwide longevities.

Global inequalities of all kinds are not going to be eliminated in a generation or two, and hence a realistic goal is not any rapid converging toward an egalitarian consumption mean: that mean would require significant consumption cuts in some of the richest countries (halving today's average per capita supply) and some substantial increases in the poorest ones (doubling today's per capita availability). What is desirable and what should be pursued by all possible means is a gradual convergence toward that egalitarian mean combined with continuing efficiency improvements and with practical displacement of some meat consumption by environmentally less demanding animal foodstuffs.

Such a process would be benefiting everybody by improving health and life expectancies of both affluent and low-income populations and by reducing the environmental burdens of meat production. Although the two opposite consumption trends of this great transition have been evident during the past generation, a much less uneven distribution of meat supply could come about only as a result of complex adjustments that will take decades to unfold. In the absence of dietary taboos, average meat intakes can rise fast as disposable incomes go up; in contrast, food preferences are among the most inertial of all behavioral traits and (except as result of a sudden economic hardship) consumption cuts of a similar rapidity are much less likely.

At the same time, modern dietary transition has modified eating habits of most of the humanity in what have been, in historic terms, relative short spans of time, in some cases as brief as a single generation. These dietary changes have been just a part of the general post-WW II shift toward greater affluence, and the two generations of these (only mildly interrupted) gains have created a habit of powerful anticipations of further gains. That may not be the case during the coming two generations because several concatenated trends are creating a world that will be appreciably different from that whose apogee was reached during the last decade of the 20th century.

Aging of Western population and, in many cases, their absolute decline appear to be irreversible processes: fertilities have fallen too far to recover above the replacement level, marriage rates are falling, first births are being postponed while the cost of raising a family in modern cities has risen considerably. By 2050, roughly two out of five Japanese, Spaniards and Germans will be above 60 years of age; even in China that share will be one-third (compared to just 12% in 2010!), and, together with many smaller countries, Germany, Japan and Russia will have millions (even tens of millions) fewer people than they have today.

We have yet to understand the complex impacts of these fundamental realities, but (judging by the German, Japanese and even Chinese experiences) continuing rise in meat demand will not be one of them. And while the American population will continue to grow, the country's extraordinarily high rate of overweight and obesity, accompanied by a no less extraordinary waste of food, offer a perfect justification for greatly reduced meat consumption. Beef consumption is already in long-term decline, and the easiest way to achieve gradual lowering of America's overall per capita meat intakes would not be by appealing to environmental consciousness (or by pointing out exaggerated threats to health) but by paying a price that more accurately reflects meat's claim on energy, soils, water and the atmosphere.

Meat, of course, is not unique as we do not pay directly for the real cost of any foodstuff we consume or any form of energy that powers the modern civilizations or raw material that makes its complex infrastructures. Meat has become more affordable not only because of the rising productivity of the livestock sector but also because much less has been spent on other foodstuffs. This post-WW II spending shift has been pronounced even in the US where food was already abundant and relatively inexpensive: food expenditures took more than 40% of an average household's disposable income in 1900; by 1950, the share was about 21%; it fell below 15% in 1966 and below 10% (9.9%) in the year 2000; in 2010, it was 9.4%, with just 5.5% spent on food consumed at home and 3.9% on food eaten away from home (USDA 2012b). The total expenditure was slightly less than spending on recreation and much less than spending on health care. At the same time, the share of overall food and drink spending received by farmers shrank from 14% in 1967 to 5% in 2007, while the share going to restaurants rose from 8% to 14%.

These trends cannot continue, and their arrest and a partial reversal should be a part of the affluent world's broader return to rational spending after decades of living beyond its means. Unfortunately, such adjustments may not be gradual: while the FAO food price index stayed fairly steady between 1990 and 2005, the post-2008 spike lifted it to more than double the 2002–2004 mean, and it led to renewed concerns about future food supply and about the chances of recurring, and even higher, price spikes. Increased food prices in affluent countries would undoubtedly reduce the overall meat consumption, but their effect on food security on low-income nations is much less clear. For decades, low international food prices were seen as a major reason for continuing insecurity of their food supply (making it impossible for small-scale farmers to compete), but that conclusion was swiftly reversed with the post-2007 rapid rise of commodity prices that came to be seen as a major factor pushing people into hunger and poverty (Swinnen and Squicciarini 2012).

In any case, it is most unlikely that food prices in populous nations of Asia and Africa will decline to levels now prevailing in the West: China's share of food spending is still 25% of disposable income, and given the country's chronic water shortages, declining availability of high-quality farmland and rising feed imports, it is certain that it will not be halved yet again by the 2030s as it was during the past generation. And the food production and supply situation in India, Indonesia, Pakistan, Nigeria or Ethiopia is far behind China's achievements, and it will put even greater limits on the eventual rise in meat demand. In a rational world, consumers in the rich countries should be willing to pay more for a food in order to lower the environmental impacts of its production, especially when that higher cost and the resulting lower consumption would also improve agriculture's long-term prospects and benefit the health of the affected population.

So far, modern societies have shown little inclination to follow such a course – but I think that during the coming decades, a combination of economic and environmental realities will hasten such rational changes. Short-term outlook for complex systems is usually more of the same, but (as in the past) unpredictable events (or events whose eventual occurrence is widely anticipated but whose timing is beyond our ken) will eventually lead to some relatively rapid changes. These realities make it impossible to predict the durability of specific trends, but I think that during the next two to four decades, the odds are more than even that many rational adjustments needed to moderate livestock's environmental impact (changes ranging from higher meat prices and reduced meat intakes to steps leading to lower environmental impacts of livestock production) will take place – if not by design, then by the force of changing circumstances.

Most nations in the West, as well as Japan, have already seen saturations of per capita meat consumption: inexorably, growth curves have entered the last, plateauing, stage and in some cases have gone beyond it, resulting in actual consumption declines. Most low-income countries are still at various points along the rapidly ascending phase of their consumption growth curves, but some are already approaching the upper bend. There is a high probability that by the middle of the 21st century, global meat production will cease to pose a steadily growing threat to the biosphere's integrity.

References

a Calorie Counter. 2012. *Fast Food Restaurants & Nutrition Facts Compared.* http://www.acaloriecounter.com/fast-food.php. Accessed on November 14, 2012.

Aarestrup, F. 2012. Get pigs off antibiotics. *Nature* 486:465–466.

Abel, W. 1962. *Geschichte der deutschen Landwirtschaft von frühen Mittelalter bis zum 19 Jahrhundert.* Stuttgart: Ulmer.

Aberle, E.D. et al. 2001. *Principles of Meat Science.* Dubuque, IA: Kendall Hunt Publishing.

Adams, C. 2010. *The Sexual Politics of Meat: A Feminist-Vegetarian Critical Theory.* New York: Continuum International Publishing Group. http://books.google.ca/books/about/The_Sexual_Politics_of_Meat.html?id=AwrwRKNavtAC. Accessed on November 14, 2012.

AFSSA (*Agence française_de_sécurité* sanitaire des produits de santé). 2009. *Synthèse de l'étude Individuelle Nationale des Consommations Alimentaires 2 (INCA 2) 2006–2007.* Paris: AFSSA.

Aiello, L.C. and J.C.K. Wells. 2002. Energetics and the evolution of the genus *Homo. Annual Review of Anthropology* 31:323–338.

Aiello, L.C. and P. Wheeler. 1995. The expensive-tissue hypothesis. *Current Anthropology* 36:199–221.

Al-Deseit, B. 2009. Least-cost broiler ration formulation using linear programming technique. *Journal of Animal and Veterinary Advances* 8:1274–1278.

Allan, J.A. 1993. Fortunately there are substitutes for water otherwise our hydro-political futures would be impossible. In: *Priorities for Water Resources Allocation and Management.* London: ODA, pp. 13–26.

Allbaugh, L.G. 1953. *Crete: A Case Study of an Undeveloped Area.* Princeton, NJ: Princeton University Press.

Allen, R.C. 2007. *How Prosperous Were the Romans? Evidence from Diocletian's Price Edict (301 AD).* Oxford: Department of Economics, Oxford University.

Should We Eat Meat?: Evolution and Consequences of Modern Carnivory,
First Edition. Vaclav Smil.
© 2013 John Wiley & Sons, Ltd. Published 2013 by John Wiley & Sons, Ltd.

Allen, J.R.M. 2010. Last glacial vegetation of northern Eurasia. *Quaternary Science Reviews* 29:2604–2618.

de Almeida, J.C. et al. 2006. Fatty acid composition and cholesterol content of beef and chicken meat in Southern Brazil. *Brazilian Journal of Pharmaceutical Sciences* 42:109–117.

Alonso, A. et al. 2009. Cardiovascular risk factors and dementia mortality: 40 years of follow-up in the Seven Countries Study. *Journal of Neurological Sciences* 2009:79–83.

Alroy, J. 2001. A multispecies overkill simulation of the end-Pleistocene megafaunal mass extinction. *Science* 292:1893–1896.

Alvard, M.S. and L. Kuznar. 2001. Deferred harvests: The transition from hunting to animal husbandry. *American Anthropologist* 103:295–311.

Alverson, D.L. 2005. Managing the catch of non-target species. In: W.S. Wooster and J.M. Quinn, eds., *Improving Fishery Management: Melding Science and Governance*. Seattle, WA: The School of Marine Affairs, University of Washington.

Alverson, D.L. et al. 2004. *A Global Assessment of Fisheries Bycatch and Discards*. Rome: FAO.

AMI (American Meat Institute). 2012. *The United States Meat Industry at a Glance*. http://www.meatami.com/ht/d/sp/i/47465/pid/47465. Accessed on November 14, 2012.

Amit, M. 2010. Vegetarian diets in children and adolescents. *Pediatric Child Health* 15(3):303–308.

Anderson, O.E. 1953. *Refrigeration in America: A History of a New Technology and Its Impact*. Princeton, NJ: Princeton University Press.

Animal Liberation Front. 2012. *Manifesto for Radical Abolitionism: By Any Means*. http://www.animalliberationfront.com/ALFront/Manifesto-TotalLib.htm. Accessed on November 14, 2012.

Antony, A.C. 2003. Vegetarianism and vitamin B-12 (cobalamin) deficiency. *American Journal of Clinical Nutrition* 78:3–6.

Århem, K. 1989. Maasai food symbolism: The cultural connotations of milk, meat, and blood in the pastoral Maasai diet. *Anthropos* 8:1–23.

Armelagos, G.J. and K.N. Harper. 2005. Genomics at the origins of agriculture, part one. *Evolutionary Anthropology* 14:68–77.

Asner, G.P. et al. 2004. Grazing systems, ecosystem responses and global change. *Annual Review of Environment and Resources* 29:261–299.

Atalay, S. and C.A. Hastorf. 2006. Food, meals, and daily activities: Food *habitus* at Neolithic Çatalhöyük. *American Antiquity* 71:283–319.

Atwater, W.O. 1888. Foods and beverages. *The Century Magazine*. May, pp. 135–139.

Atwater, W. and C. Woods. 1896. *The Chemical Composition of American Food Materials*. Washington, DC: USDA.

Australian Government. 2012. *Australian Standards for the Export of Livestock (Version 2.3)*. http://www.daff.gov.au/animal-plant-health/welfare/export-trade/livestock-export-standards. Accessed on November 14, 2012.

Avery, A. and D. Avery. 2007. *The Environmental Safety and Benefits of Growth Enhancing Pharmaceutical Technologies in Beef Production*. Washington, DC: Hudson Institute. http://www.cgfi.org/pdfs/nofollow/beef-eco-benefits-paper.pdf. Accessed on November 14, 2012.

Aviagen. 2009. *Ross Broiler Management Manual: Chick Management*. Newbridge: Aviagen.

AWI (Animal Welfare Institute). 2010. *The Welfare of Chickens Raised for Meat*. Washington, DC: AWI.

Bailey, L.H., ed. 1908. *Cyclopedia of American Agriculture*. New York: Macmillan.

Bailey, R.C. and T.N. Headland. 1991. The tropical rain forest: Is it a productive environment for human foragers? *Human Ecology* 19:261–285.

Bailey, R.C. et al. 1989. Hunting and gathering in tropical rain forest: Is it possible? *American Anthropologist* 91:59–82.

Ball, C.E. 1998. *Building the Beef Industry: A Century of Commitment 1898–1998*. Saratoga, WY: The Saratoga Publishing Group.

Barbosa, P.M., D. Stroppiana, and J.-M. Grégoire. 1999. An assessment of vegetation fire in Africa (1981–1991): Burned areas, burned biomass, and atmospheric emissions. *Global Biogeochemical Cycles* 13:933–949.

Barles, S. 2007. Feeding the city: Food consumption and flow of nitrogen, Paris, 1801–1914. *Science of the Total Environment* 375:48–58.

Bar-Yosef, O. 2002. The Upper Paleolithic Revolution. *Annual Review of Anthropology* 31:363–393.

Becker, J. 1998. *Hungry Ghosts: Mao's Secret Famine*. New York: Holt.

Beckett, J.L. and J.W. Oltjen. 1993. Estimation of the water requirement for beef production in the United States. *Journal of Animal Science* 71: 818–826.

Belcher, J.N. 2006. Industrial packaging development for the global meat market. *Meat Science* 74:143–148.

Bender, A. 1992. *Meat and Meat Products in Human Nutrition in Developing Countries*. Rome: FAO.

Bengoa, J.M. 2001. Food transition in the 20th–21st century. *Public Health Nutrition* 4(6A):1425–1427.

deBenoist, B. et al. 2008. *Worldwide Prevalence of Anaemia 1993–2005*. Geneva: WHO. http://whqlibdoc.who.int/publications/2008/9789241596657_eng.pdf. Accessed on November 14, 2012.

Bentley, A. 1998. *Eating for Victory: Food Rationing and the Politics of Domesticity*. Urbana, IL: University of Illinois Press.

Berley, P. and Z. Singer. 2007. *The Flexitarian Table: Inspired, Flexible Meals for Vegetarians, Meat Lovers, and Everyone in Between*. New York: Houghton Mifflin Harcourt.

Beyond Meat™. 2012. *More Please*. http://beyondmeat.com/. Accessed on November 14, 2012.

Bhattacharya, A. et al. 2006. Biological effects of conjugated linoleic acid in health and disease. *Journal of Nutritional Biochemistry* 17:789–810.

Bifarin, J.O., M.E. Ajibola, and A.A. Fadiyimu. 2008. Analysis of marketing bush meat in Idanre Local Government area of Ondo state, Nigeria. *African Journal of Agricultural Research* 3:667–671.

Biligilli, S.F. and J.B. Hess. 1995. Placement density influences broiler carcass grade and meat yields. *Journal of Applied Poultry Research* 4:384–389.

Binford, L.R. 1981. *Bones: Ancient Men and Modern Myths*. New York: Academic Press.

Bird Life International. 2011. *UPDATE: Over 866,000 Birds Slaughtered So Far This Autumn in Cyprus! – Sign the Petition to Stop This Now!* Thursday, October 13. http://www.birdlife.org/community/2011/10/update-over-866000-birds-slaughtered-so-far-this-autumn-in-cyprus-sign-the-petition-to-stop-this-now/. Accessed on November 14, 2012.

Blackmore, D.K. 1993. Euthanasia; not always Eu. *Australian Veterinary Journal* 70:409–412.

Blascoa, R., J.P. Peris, and J. Rosella. 2010. Several different strategies for obtaining animal resources in the late Middle Pleistocene: The case of level XII at Bolomor Cave (Valencia, Spain). *Comptes Rendus Palevol* 9:171–184.

Bleken, M.A. and L.R. Bakken. 1997. Nitrogen cost of food production: Norwegian Society. *AMBIO* 26:134–142.

Block, B.A. 1994. Thermogenesis in muscle. *Annual Review of Physiology* 56: 535–577.

BLS (Bureau of Labor Statistics). 2012. *Incidence Rate and Number of Nonfatal Occupational Injuries by Industry and Ownership, 2010*. http://www.bls.gov/iif/oshwc/osh/os/ostb2805.pdf. Accessed on November 14, 2012.

Blumenschine, R.J. 1991. Hominid carnivory and foraging strategies, and the socio-economic function of early archaeological sites. *Philosophical Transactions of the Royal Society* 334:211–221.

Blumenschine, R.J. and J.A. Cavallo. 1992. Scavenging and human evolution. *Scientific American* 267(4):90–95.

Boesch, C. 1994. Chimpanzees – Red colobus: A predator-prey system. *Animal Behaviour* 47:1135–1148.

Boesch, C. and H. Boesch-Achermann. 2000. *The Chimpanzees of the Taï Forest*. Oxford: Oxford University Press.

Boyd, F. 1994. Humane slaughter of poultry: The case against the use of electrical stunning devices. *Journal of Agricultural and Environmental Ethics* 7:221–236.

Boyd, W. 2003. Making meat: Science, technology, and American poultry production. *Technology and Culture* 42:631–664.

Boyd, W. and M. Watts. 1997. Agro-industrial just-in-time: The chicken industry and postwar American capitalism. In: D. Goodman and M. Watts, eds., *Globalising Food: Agrarian Question and Global Restructuring*. New York: Routledge, pp. 192–225.

Boyle, E. 2012. *High Steaks: Why and How to Eat Less Meat*. Gabriola Island, BC, Canada: New Society Publishers.

Bradford, E. et al. 1999. *Animal Agriculture and Global Food Supply*. Ames, IO: Council for Agricultural Science and Technology.

Braun, D.R. et al. 2010. Early hominin diet included diverse terrestrial and aquatic animals 1.95 Ma in East Turkana, Kenya. *Proceedings of National Academy of Sciences* 107:10002–10007.

Britton, D. 1999. Meat is madness. *Times Literary Supplement.* September 17, p. 29.

Brown, E. 2009. *The Meatball Cookbook Bible: Foods from Soups to Deserts-500 Recipes that Make the World Go Round.* Kennebunkport, ME: Cider Mill Press.

Bryan, N.S. 2011. Letter by Bryan regarding article, 'Red and processed meat consumption and risk of incident coronary heart disease, stroke, and diabetes mellitus'. *Circulation* 123:e16.

Buck, J.L. 1930. *Chinese Farm Economy.* Nanjing: University of Nanking.

Buck, J.L. 1937. *Land Utilization in China.* Nanjing: University of Nanking.

Butler, D. 2012. Death-rate row blurs mutant flu debate. *Nature* 482:289.

Caballero, B. and B.M. Popkin, eds. 2002. *The Nutrition Transition: Diet and Disease in the Developing World.* Amsterdam: Academic Press.

Cachel, S. 1997. Dietary shifts and the European Upper Paleolithic transition. *Current Anthropology* 38:579–603.

Calverd, S. 2005. A radical approach to Kyoto. *Physics World.* July, p. 56.

Capper, J.L., R.A. Cady, and D.E. Bauman. 2011. *Demystifying the Environmental Sustainability of Food Production.* http://txanc.org/wp-content/uploads/2011/08/8_Capper_Demystifying-the-Environmental-Sustain-of-Food-Prod_FINAL.pdf. Accessed on November 14, 2012.

CARC (Canadian Agri-Food Research Council). 2003. *Recommended Code of Practice for the Care and Handling of Farm Animals: Chickens, Turkeys and Breeders from Hatchery to Processing Plant.* Ottawa, ON, Canada: CARC.

Cavalieri, P. 2003. *The Animal Question: Why Non-Human Animals Deserve Human Rights.* Oxford: Oxford University Press.

CDCP. 2011. *National Health and Nutrition Examination Survey.* http://www.cdc.gov/nchs/nhanes/nhanes2007-2008/nhanes07_08.htm. Accessed on November 14, 2012.

CDCP. 2012. Escherichia coli *O157:H7 and Other Shiga Toxin Producing Escherichia coli (STEC).* http://www.cdc.gov/nczved/divisions/dfbmd/diseases/ecoli_o157h7/#how_common. Accessed on November 14, 2012.

Cease, A.J. et al. 2012. Heavy livestock grazing promotes locust outbreaks by lowering plant nitrogen content. *Science* 335:467–469.

CEU (Council of the European Union). 1993. *Council Directive 93/119/EC of 22 December 1993 on the Protection of Animals at the Time of Slaughter or Killing.* http://ec.europa.eu/food/fs/aw/aw_legislation/slaughter/93-119-ec_en.pdf. Accessed on November 14, 2012.

Chaker, A.M. 2012. The veggie burger's new dream: Be more like meat. *The Wall Street Journal.* January 25.

Chang, K.C., ed. 1977. *Food in China's Culture.* New Haven, CT: Yale University Press.

Chapagain, A.K. and A.Y. Hoekstra. 2004. *Water Footprints of Nations, Volume 1: Main Report.* Paris: UNESCO-IHE.

Chapin, F.S., P. Matson, and H.A. Mooney. 2002. *Principles of Terrestrial Ecosystem Ecology*. New York: Springer-Verlag.

Chapple, C. 1993. *Nonviolence to Animals, Earth and Self in Asian Traditions*. Albany, NY: State University of New York Press.

Charbonneau, R. 1988. Fiesta for six: One guinea pig…and we'll all be full. *IDRC Reports*. July, pp. 6–8.

Chen, H. et al. 2004. The evolution of H5N1 influenza viruses in ducks in southern China. *Proceedings of the National Academy of Sciences* 101:10452–10457.

Childe, V.G. 1951. The Neolithic revolution. In: V.G. Childe, ed., *Man Makes Himself*. London: C.A. Watts, pp. 67–72.

Chivers, D.J. and C.M. Hladik. 1980. Morphology of the gastrointestinal tract in primates: Comparisons with other mammals in relation to diet. *Journal of Morphology* 166:337–386.

Chotpitayasunondh, T. et al. 2004. Human disease from influenza A (H5N1), Thailand, 2004. *Emerging Infectious Diseases* 11:201–209.

Cipriani, L. 1962. *California and Overland Diaries of Count Leonetto Cipriani from 1853 Through 1871*. Portland, OR: Champoeg Press.

CIV (Centre d'Information des Viandes). 2009. *L'alimentation des Français*. http://www.civ-viande.org/file/documents/dossier-sante-alimentation-francais.pdf. Accessed on November 14, 2012.

CIV. 2012. *Meat Conversion Ratios*. http://www.civ-viande.org/uk/ebn.ebn?pid=57&rubrik=6&item=36&page=77. Accessed on November 14, 2012.

Clark, G., M. Huberman, and P. Lindert. 1995. A British food puzzle, 1770–1850. *Economic History Review* 48:215–237.

Clark, M. and S. Spaull. 2010. *Leiths Meat Bible*. New York: Bloomsbury.

Clastres, P. 1981. *Chronicle of the Guayaki Indians: The Aché, Nomadic Hunters of Paraguay*. New York: Urizen Books.

Clay, J. 2011. Freeze the footprint of food. *Nature* 475:287–289.

Clutton-Brock, J. 1999. *A Natural History of Domesticated Mammals*. Cambridge: Cambridge University Press.

CME (Chicago Mercantile Exchange). 2004. *Introduction to Livestock and Meat Fundamentals*. Chicago, IL: CME.

Cobbett, W. 1824. *Cottage Economy*. New York: Gould.

Codazzi, D. et al. 2005. Coma and respiratory failure in a child with severe vitamin B_{12} deficiency. *Pediatric Critical Care and Medicine* 6:483–485.

Coe, M.J., D.H. Cumming, and J. Phillipson. 1976. Biomass and production of large African herbivores in relation to rainfall and primary production. *Oecologia* 22:341–354.

Cohen, M.N. 2000. History, diet, and hunter-gatherers. In: K.F. Kiple and K.C. Ornelas, eds., *The Cambridge World History of Food*. Cambridge: Cambridge University Press, pp. 63–69.

Colby, H. and J. Greene. 1998. Statistical revision significantly alters China's livestock PS&D. *FAS Online*. October 23. http://www.fas.usda.gov. Accessed on November 14, 2012.

Connor, M.L. et al. 1993. *Recommended Code of Practice for the Care and Handling of Farm Animals: Pigs*. Ottawa, ON, Canada: Agriculture and Agri-Food Canada.

Cordain, L. 2012. *The Paleo Diet Cookbook: More Than 150 Recipes for Paleo Breakfasts, Lunches, Dinners, Snacks and Beverages*. New York: John Wiley & Sons.

Cordain, L. et al. 2000. Plant-animal subsistence ratios and macronutrient energy estimations in worldwide hunter-gatherer diets. *American Journal of Clinical Nutrition* 71:682–692.

Cordain, L. et al. 2002. The paradoxical nature of hunter-gatherer diets: Meat-base, yet not-atherogenic. *European Journal of Clinical Nutrition* 56 (Suppl. 1):542–552.

Corpet, D.E. 2011. Red meat and cancer: Should we become vegetarians, or can we make meat safer? *Meat Science* 89:310–316.

Couch, P. 2012. *Grilling Gone Wild: Zesty Recipes for Meats, Mains, Marinades and More*. East Petersburg, PA: Fox Chapel Publishing.

Craig, O.E. et al. 2010. Stable isotope analysis of Late Upper Palaeolithic human and animal remains from Grota del Romito (Cosenza), Italy. *Journal of Archaeological Science* 37:2504–2512.

Crawley, M.J. 1983. *Herbivory: The Dynamics of Animal-Plant Interactions*. Berkeley, CA: University of California Press.

Critchell, J.T. and J. Raymond. 1912. *A History of the Frozen Meat Trade*. London: Constable & Company.

Crutzen, P. et al. 1986. Methane production by domestic animals, wild ruminants, other herbivorous fauna, and humans. *Tellus* 38B:271–284.

Cui, J., Z.Q. Wang, and B.L. Xu. 2011. The epidemiology of human trichinllosis in China during 2004–2009. *Acta Tropica* 118:1–5.

Damuth, J. 1981. Population density and body size in mammals. *Nature* 290: 699–700.

Daniel, C.R. et al. 2011. Trends in meat consumption in the United States. *Public Health and Nutrition* 14:575–583.

Dardenne, E. 2010. The reception of Peter Singer's theories in France. *Society and Animals* 18:205–218.

Darmon, N. and A. Drewnowski. 2008. Does social class predict diet quality? *American Journal of Clinical Nutrition* 87:1107–1117.

Datar, I. and M. Betti. 2010. Possibilities for an in vitro meat production system. *Innovative Food Science and Emerging Technologies* 11:13–22.

Davidson, E.A. et al. 2012. The Amazon basin in transition. *Nature* 481:321–327.

Dawson, M. 2009. *Plenti and Grase: Food and Drink in the Sixteenth-Century Household*. Totnes, Devon: Prospect.

DeGrazia, D. 1996. *Taking Animals Seriously: Mental Life and Moral Status*. Cambridge: Cambridge University Press.

Derven, D. 1984. Deerfield foodways. In: P. Benes and J. Montague, eds., *Foodways in the Northeast*. Boston, MA: Boston University, pp. 47–63.

Diener, P. and E.E. Robkin. 1978. Ecology, evolution, and the search for cultural origins: The question of Islamic pig prohibition. *Current Anthropology* 19:494–509.

Digard, J.-P. 1990. *L'Homme et les animaux domestiques: Anthropologie d'une passion*. Paris: Fayard.

Domínguez-Rodrigo, M. 2002. Hunting and scavenging by early humans: The state of the debate. *Journal of World Prehistory* 16:1–54.

Domínguez-Rodrigo, M. and T.R. Pickering. 2003. Early hominid hunting and scavenging: A zooarcheological review. *Evolutionary Anthropology* 12: 275–282.

Domínguez-Rodrigo, M. et al. 2009. Unraveling hominin behavior at another anthropogenic site from Olduvai Gorge (Tanzania): New archaeological and taphonomic research at BK, Upper Bed II. *Journal of Human Evolution* 57:260–283.

Dregne, H.E. and N. Chou. 1994. Global desertification and costs. In: H.E. Dregne, ed., *Degradation and Restoration of Arid Lands*. Lubbock, TX: Texas Tech University, pp. 249–282.

Driscoll, C.A., D.W. Macdonald, and S.J. O'Brien. 2009. From wild animals to domestic pets, an evolutionary view of domestication. *Proceedings of the National Academy of Sciences* 106:9971–9978.

Drucker, D. and H. Bocherens. 2004. Carbon and nitrogen stable isotopes as tracers of change in diet breadth during Middle and Upper Palaeolithic in Europe. *International Journal of Osteoarchaeology* 14:162–177.

Dupin, H. et al. 1984. Evolution of the French diet: Nutritional aspects. *World Review of Nutrition and Dietetics* 44:57–84.

Dürrwächter, C. et al. 2006. Beyond the grave: Variability in Neolithic diets in Southern Germany. *Journal of Archaeological Science* 33:39–48.

Dyson-Hudson, B. 1980. Strategies of resource exploitation among East African savanna pastoralist. In: D.R. Harries, ed., *Human Ecology in Savanna Environments*. London: Academic Press, pp. 171–184.

Eaton, S.B. 1992. Humans, lipids and evolution. *Lipids* 27:814–819.

Eaton, S.B. and S.B. Eaton III. 2000. Paleolithic vs. modern diets – Selected pathophysiological implications. *European Journal of Nutrition* 39:67–70.

Eaton, S.B., S.B. Eaton, III, and M.J. Konner. 1997. Paleolithic nutrition revisited: A twelve-year retrospective on its nature and implications. *European Journal of Clinical Nutrition* 51:207–216.

Eaton, S.B. and M. Konner. 1985. Paleolithic nutrition. A consideration of its nature and current implications. *New England Journal of Medicine* 312:283–289.

EC (European Commission). 2005. *Ban on Antibiotics as Growth Promoters in Animal Feed Enters into Effect*. http://europa.eu/rapid/pressReleasesAction. do?reference=IP/05/1687&format=HTML&aged=0&language=EN#fnB1. Accessed on November 14, 2012.

Edelman, P.D. et al. 2005. In vitro-cultured meat production. *Tissue Engineering* 11:659–662.

Eden, F.M. 1797. *The State of the Poor*. London: Davis.

Elvin, M. 2004. *The Retreat of the Elephants: An Environmental History of China*. New Haven, CT: Yale University Press.

Emborg, H. et al. 2001. The effect of discontinuing the use of antimicrobial growth promoters on the productivity in the Danish broiler production. *Preventive Veterinary Medicine* 50:53–70.

Engber, D. 2008. The bogus $1 million meat prize. *Slate*. April 23. http://www.unz.org/Pub/Slate-2008apr-00317. Accessed on November 14, 2012.

Fa, J.E. and D. Brown. 2009. Impacts of hunting on mammals in African tropical moist forests: A review and synthesis. *Mammal Review* 39:231–264.

Fagone, J. 2006. Horsemen of the esophagus. *The Atlantic Monthly*. May, pp. 86–93.

Fairchild, B.D. 2005. *Broiler Stocking Density*. Athens, GA: The University of Georgia Cooperative Extension Service. http://www.poultry.uga.edu/extension/tips/documents/01%2005%20B%20tip%20B%20F%20(web).pdf. Accessed on November 14, 2012.

FAO (Food and Agriculture Organization). 2001. *Guidelines for Humane Handling, Transport and Slaughter of Livestock*. Rome: FAO.

FAO. 2004. *Protein Sources for the Animal Feed Industry*. Rome: FAO.

FAO. 2005. *Review of the State of World Marine Fishery Resources*. Rome: FAO.

FAO. 2010a. *The State of Food Insecurity in the World*. Rome: FAO.

FAO. 2010b. *Fats and Fatty Acids in Human Nutrition*. Rome: FAO.

FAO. 2010c. *Good Practices for Feed Industry*. Rome: FAO.

FAO. 2010d. *The State of World Fisheries and Aquaculture 2010*. Rome: FAO.

FAO. 2012. *FAOSTAT*. http://faostat.fao.org/. Accessed on November 14, 2012.

FAO/WHO (Food and Agriculture Organization/World Health Organization). 1993. *Energy and Protein Requirements. Report of a Joint FAO/WHO Ad hoc Expert Committee*. Rome: FAO.

FAO/WHO. 2007. *Protein and Amino Acid Requirements in Human Nutrition*. Geneva: WHO.

FDA (Food and Drug Administration). 2010. *2009 Summary Report on Antimicrobials Sold or Distributed for Use in Food-Producing Animals*. Washington, DC: FDA. http://www.fda.gov/downloads/ForIndustry/UserFees/AnimalDrugUserFeeActADUFA/UCM231851.pdf. Accessed on November 14, 2012.

Ferguson, L.R. 2010. Meat and cancer. *Meat Science* 84:308–313.

Ferguson, N.M., Donnelly, C.A., and R.M. Anderson. 2001. The foot-and-mouth epidemic in Great Britain: Pattern of spread and impact of interventions. *Science* 292:1155–1160.

Ferrières, J. 2004. The French paradox: Lessons for other countries. *Heart* 90:107–111.

Finch, C.E. and C.B. Stanford. 2004. Meat-adaptive genes and the evolution of slower aging in humans. *The Quarterly Review of Biology* 79:3–50.

Fish, J.L. and C.A. Lockwood. 2003. Dietary constraints on encephalization in primates. *American Journal of Physical Anthropology* 12:171–181.

Flachowsky, G. 2011. Carbon-footprints for food of animal origin, reduction potentials and research needs. *Journal of Applied Animal Research* 39:2–14.

Flachowsky, G. and S. Hachenberg. 2009. CO_2-footprints for food of animal origin – Present stage and open questions. *Journal für Verbraucherschutz und Lebensmittelsicherheit* 4:190–198.

Flandrin, J.-L. 1989. Distinction through taste. In: R. Chartier, ed., *A History of Private Life III: Passions of the Renaissance*. Cambridge, MA: The Belknap Press, pp. 265–307.

Flandrin, J.-L. and M. Montanari, eds. 1996. *Histoire de l'alimentation*. Paris: Edition Fayard.

FNB (Food and Nutrition Board). 2005. *Dietary Reference Intakes for Energy, Carbohydrates, Fiber, Fat, Fatty Acids, Cholesterol, Protein, and Amino Acids*. Washington, DC: The National Academies Press.

Fogel, R.W. 1991. The conquest of high mortality and hunger in Europe and America: Timing and mechanisms. In: P. Higgonet et al., eds., *Favorites of Fortune*. Cambridge, MA: Harvard University Press, pp. 33–71.

Fogel, R.W. 2004. *The Escape from Hunger and Premature Death, 1700–2100: Europe, America, and the Third World*. Cambridge: Cambridge University Press.

Foley, R.A. and P.C. Lee. 1991. Ecology and energetics of encephalization in hominid evolution. *Philosophical Transactions of the Royal Society of London B* 334:223–232.

Foley, J.A. et al. 2011. Solutions for a cultivated planet. *Nature* 478:337–342.

Ford, B.J. 2011. Impact of cultured meat on global agriculture. *World Agriculture* 2(2):43–46.

Ford, B.J. 2012. *Meat: The Story Behind Our Greatest Addiction*. Oxford: Oneworld.

Franco, M.P. et al. 2007. Human brucellosis. *Lancet Infectious Diseases* 7:775–786.

Frank, E. and R. White. 1994. Health effects and prevalence of vegetarianism. *The Western Journal of Medicine* 160:465–470.

Freeman, S.R. et al. 2009. Alternative methods for disposal of spent laying hens: Evaluation of the efficacy of grinding, mechanical deboning, and of keratinase in the rendering process. *Bioresource Technology* 100:4515–4520.

French, N.R. et al. 1976. Small mammal energetic in grassland ecosystems. *Ecological Monographs* 46:201–220.

French, P. et al. 2000. Fatty acid composition, including conjugated linoleic acid, of intramuscular fat from steers offered grazed grass, grass silage, or concentrate-based diets. *Journal of Animal Science* 78:2849–2855.

Frenzen, P.D., A. Drake, and F.J. Angulo. 2005. Economic cost of illness due to Escherichia coli O157 infections in the United States. *Journal of Food Protection* 68:2623–2630.

Frison, G.C. 1987. Prehistoric hunting strategies. In: M.H. Nitecki and D.V. Nitecki, eds., *The Evolution of Human Hunting*. New York: Plenum Press, pp. 177–223.

Fritzson, A. and T. Berntsson. 2006. Energy efficiency in the slaughter and meat processing industry – Opportunities for improvements in future energy markets. *Journal of Food Engineering* 77:792–802.

Fuller, M.F. 2004. *The Encyclopedia of Farm Animal Nutrition*. Wallingford: CABI Publishing.

Galaty, J.G. and P.C. Salzman, eds. 1981. *Change and Development in Nomadic and Pastoral Societies*. Leiden: E.J. Brill.

Gallet, C.A. 2009. Meat meets meta: A quantitative review of the price elasticity of meat. *American Journal of Agricultural Economics* 92:258–272.

Galloway, J.N. et al. 2004. Nitrogen cycles: Past, present, and future. *Biogeochemistry* 70:153–226.

Galloway, J.N. et al. 2007. International trade in meat: The tip of the pork chop. *Ambio* 36:622–629.

Garcia-Guixe, E. et al. 2009. Stable isotope analysis of human and animal remins from the Late Upper Palaeolithic site of Balma Guilanya, southeastern Pre-Pyrenees, Spain. *Journal of Archaeological Science* 36:1018–1026.

Garfunkel, T. 2004. *Kosher for Everybody: The Complete Guide to Understanding, Shopping, Cooking, and Eating the Kosher Way*. San Francisco, CA: Jossey-Bass.

Garnsey, P. 1999. *Food and Society in Classical Antiquity*. Cambridge: Cambridge University Press.

Gelfand, I., S.S. Snapp, and G.P. Robertson. 2010. Energy efficiency of conventional, organic and alternative cropping systems for food and fuel at a site in the US Midwest. *Environmental Science & Technology* 44:4006–4011.

Gerber, N. et al. 2009. The influence of cooking and fat trimming on the actual nutrient intake from meat. *Meat Science* 81:148–154.

Gerber, P. et al., eds. 2010. *Livestock in a Changing Landscape*. Washington, DC: Island Press.

Giannini, A. et al. 2006. Health risk for children on vegan or vegetarian diets. *Pediatric Critical Care and Medicine* 7:188.

Gilbert, N. 2012. Rules tighten on use of antibiotics on farms. *Nature* 481:125.

Gill, C. 2006. Feed more profitable, but disease breeds uncertainty. *Feed International*. January, pp. 5–11. http://s3.amazonaws.com/zanran_storage/www.ifif.org/ContentPages/2470468148.pdf. Accessed on November 14, 2012.

Ginter, E. 2008. Vegetarian diets, chronic diseases and longevity. *Bratislavske Lekarske Listy* 109:463–466.

Givens, D.I. 2010. Milk and meat in our diet: Good or bad for health? *Animal* 4:1941–1952.

Gomes, C.M. and C. Boesch. 2009. Wild chimpanzees exchange meat for sex on a long-term basis. *PLoS ONE* 4(4):e5116.

Goodland, R. and J. Anhang. 2009. Livestock and climate change. *World Watch*. November/December, pp. 10–19.

Goren-Inbar, N. et al. 2004. Evidence of hominin control of fire at Gesher Benot Ya`aqov, Israel. *Science* 304:725–727.

Goudsblom, J. 1992. *Fire and Civilization*. London: Allen Lane.

Grandin, T. 2000. *Livestock Handling and Transport*. Wallingford: CABI Publishing.

Grandin, T. 2002. Return-to-sensibility problems after penetrating captive bolt stunning of cattle in commercial beef slaughter plants. *Journal American Veterinary Medical Association* 22:1258–1261.

Grandin, T. 2006. Progress and challenges in animal handling and slaughter in the U.S. *Applied Animal Behaviour Science* 100:129–139.

Grandin, T. and M. Deesing. 2008. *Humane Livestock Handling*. North Adams, MA: Storey Publishing.

Gratzer, W. 2005. *Terrors of the Table: The Curious History of Nutrition*. Oxford: Oxford University Press.

Green, A. 2005. *Filed Guide to Meat*. Philadelphia, PA: Quirk Books.

Grigg, D. 1995. The nutritional transition in Western Europe. *Journal of Historical Geography* 22:247–261.

Grivetti, L.E. 2000. Food prejudices and Taboos. In: K.F. Kiple and K.C. Ornelas, eds., *The Cambridge World History of Food*. Cambridge: Cambridge University Press, pp. 1495–1513.

Grumett, D. and R. Muers. 2010. *Theology on the Menu: Asceticism, Meat and Christian Diet*. London: Routledge.

Gustavsson, J. et al. 2011. *Global Food Losses and Food Waste*. Rome: FAO.

Guthrie, R.D. 2006. New carbon dates link climatic change with human colonization and Pleistocene extinctions. *Nature* 441:207–209.

Hairston, N. et al. 1960. Community structure, population control and competition. *The American Naturalist* 94:421–425.

Hall, K.D. et al. 2009. The progressive increase of food waste in America and its environmental impact. *PLoS ONE* 4(11):e7940.

Hanson, D.J., K. Kirchofer, and C.R. Calkins. 2001. The role of muscle glycogen in dark cutting beef. *Nebraska Beef Cattle Reports 298*. http://digitalcommons.unl.edu/animalscinbcr/298. Accessed on November 14, 2012.

Harako, R. 1981. The cultural ecology of hunting behavior among Mbuti Pygmies of the Ituri Forest, Zaire. In: R.S.O. Harding and G. Teleki, eds., *Omnivorous Primates*. New York: Columbia University Press, pp. 499–555.

Harris, M. 1966. The cultural ecology of India's sacred cattle. *Current Anthropology* 7:51–66.

Harris, M. 1974. *Cows, Pigs, Wars, and Witches: The Riddles of Culture*. New York: Random House.

Hart, E.B., J.G. Halpin, and H. Steenbock. 1920. The nutritional requirements of baby chicks. *The Journal of Biological Chemistry* 52:379–386.

Hartmann, F. 1923. *L'agriculture dans l'ancienne Egypte*. Paris: Libraire-Imprimerie Réunies.

Harvey, B. 1995. *Living and Dying in England 1100–1540, The Monastic Experience*. Oxford: Oxford University Press.

Havenstein, G.B. 2006. Performance changes in poultry and livestock following 50 years of genetic selection. *Lohmann Information* 41:30–37.

Hawkes, K., J.F. O'Connell and N.G.B. Jones. 2001. Hadza meat sharing. *Evolution and Human Behavior* 22:113–142.

Hayden, B. 1981. Subsistence and ecological adaptations of modern hunter/gatherers. In: R.S.O. Harding and G. Teleki, eds., *Omnivorous Primates.* New York: Columbia University Press, pp. 344–421.

Hébel, P. 2012. *Evolution de la consommation de viande en France: Les nouvelles données de l'enquête "Comportements et Consommations Alimentaires en France" (CCAF) 2010.* Paris: CIV. http://www.credoc.fr/pdf/Sou/Consommation_viande_CCAF2010.pdf. Accessed on November 14, 2012.

Hedges, R.E.M. and L.M. Reynard. 2007. Nitrogen isotopes and the trophic level of humans in archaeology. *Journal of Archaeological Science* 34:1240–1251.

Heichel, G.H. and N.P. Martin. 1980. Alfalfa. In: D. Pimentel, ed., *Handbook of Energy Utilization in Agriculture.* Boca Raton, FL: CRC Press, pp. 155–161.

Heinz, G. and P. Hautzinger. 2007. *Meat Processing Technology.* Rome: FAO.

Henderson, L., J. Gregory, and G. Swan. 2002. *The National Diet & Nutrition Survey: Adults Aged 19 to 64 Years.* London: HMSO.

Hennessy, D. et al. 2004. *Infectious Disease, Productivity, and Scale in Open and Closed Animal Production Systems.* Ames, IA: Center for Agricultural and Rural Development, Working Paper 04-WP 367.

Herrscher, E. and G. Le Bras-Goude. 2010. Southern French Neolithic populations: Isotopic evidence for regional specificities in environment and diet. *American Journal of Physical Anthropology* 142:259–272.

Hertrampf, J.W. 2004. Das "Wustenschiff" als Fleischlieferant: Eine Betrachtung zum Verzehr von Kamelfleisch. *Fleischwirtschaft* 84(12):111–114.

Heys, M. et al. 2010. Is childhood meat eating associated with better later adulthood cognition in a developing population? *European Journal of Epidemiology* 25:507–516.

Hilimire, K. 2011. Integrated crop/livestock agriculture in the United States: A review. *Journal of Sustainable Agriculture* 35:376–393.

Hill, K. and H. Kaplan. 1993. On why male foragers hunt and share food. *Current Anthropology* 34:701–706.

Hockett, B. and J. Haws. 2003. Nutritional ecology and diachronic trends in Paleolithic diet and health. *Evolutionary Anthropology* 12:211–216.

Hockstad, L. and Weitz, M. 2009. *Inventory of U.S. Greenhouse Gas Emissions and Sinks: 1990–2007.* Washington, DC: US Environmental Protection Agency.

Hoek, A.C. et al. 2004. Food-related lifestyle and health attitudes of Dutch vegetarians, non-vegetarian consumers of meat substitutes, and meat consumers. *Appetite* 42:265–272.

Hoek, A.C. et al. 2011. Replacement of meat by meat substitutes. A survey on person- and product-related factors in consumer acceptance. *Appetite* 56:662–673.

Hoekstra, A.Y. and A.K. Chapagain. 2007. Water footprints of nations: Water use by people as a function of their consumption pattern. *Water Resources Management* 21:35–48.

Hoering, H. and D. Chapman. 2006. *Nitrate and Nitrite in Drinking Water.* London: IWA Publishing.

Hohmann, G. and B. Fruth. 2008. Capture and meat eating by bonobos at Lui Kotale, Salonga National Park, Democratic Republic of Congo. *Folia Primatologica* 79:103–110.

Hollingsworth, D.F. 1983. Rationing and economic constraints on food consumption in Britain since the Second World War. *World Review of Nutrition and Dietetics* 42:191–218.

Holmern, T., J. Muya, and E. Røskaft. 2008. Local law enforcement and illegal bushmeat hunting outside the Serengeti. *Environmental Conservation* 34:55–63.

Holt, B.M. and V. Formicola. 2008. Hunters of the Ice Age: The biology of Upper Paleolithic people. *Yearbook of Physical Anthropology* 51:70–99.

Ho-Pham, L.T. et al. 2012. Vegetarianism, bon loss, fracture and vitamin D: A longitudinal study in Asian vegan and non-vegans. *European Journal of Clinical Nutrition* 66:75–82.

Hopper, G.R. 1999. Changing food production and quality of diet in India, 1947–98. *Population and Development Review* 25:443–477.

Horowitz, R. 2006. *Putting Meat on the American Table.* Baltimore, MD: Johns Hopkins University Press.

Hui, Y.H., ed. 1993. *Dairy Science and Technology Handbook 2 Product Manufacturing.* New York: VCH.

Hurnik, F. et al. 1991. *Recommended Code of Practice for the Care and Handling of Farm Animals: Beef Cattle.* Ottawa, ON, Canada: Agriculture and Agri-Food Canada.

HYDE. 2012. *History Database of the Global Environment.* The Hague: The Netherlands Environmental Assessment Agency. http://themasites.pbl.nl/tridion/en/themasites/hyde/. Accessed on November 29, 2012.

IFIF (International Feed Industry Federation). 2012. *Global Feed Production.* http://www.ifif.org/pages/t/Global+feed+production. Accessed on November 14, 2012.

International Zinc Consultative Group. 2004. Assessment of the risk of zinc deficiency in populations and options for its control. *Food and Nutrition Bulletin* 25:S99–S203.

IPCC (Intergovernmental Panel on Climate Change). 2006. Emissions from livestock and manure management. In: *2006 IPCC Guidelines for National Greenhouse Gas Inventories.* http://www.ipcc-nggip.iges.or.jp/public/2006gl/pdf/4_Volume4/V4_10_Ch10_Livestock.pdf. Accessed on November 14, 2012.

Ishige, N. 2001. *The History and Culture of Japanese Food.* London: Kegan Paul.

IST (Iowa State University). 2002. *Iowa Concentrated Animal Feeding Operations Air Quality Study.* Ames, IO: Iowa State University. http://www.public-health.uiowa.edu/ehsrc/CAFOstudy/CAFO_final2-14.pdf. Accessed on November 14, 2012.

Jansen, T. et al. 2002. Mitochondrial DNA and the origins of the domestic horse. *Proceedings of the National Academy of Sciences* 99:10905–10910.

Jentsch, W., B. Piatkowski, and M. Derno. 2009. Relationship between carbon dioxide production and performance in cattle and pigs. *Archiv Tierzucht* 52:485–496.

Jin, S.K. et al. 2011. The development of imitation crab sticks by substituting spent laying hen meat for Alaska Pollack. *Poultry Science* 90:1799–1808.

Jones, R. 2001. *Farm Management Guide: Finishing Beef.* Manhattan, KS: Kansas State University Agricultural Experiment Station and Cooperative Extension Service.

Jones, D. 2004. Crimes unseen. *Orion.* July/August, pp. 60–67.

Jungbluth, N. et al. 2000. Food purchases: Impacts from the consumers' point of view investigated with a modular LCA. *The International Journal of Life Cycle Assessment* 5:134–142.

Kahn, H.A. et al. 1984. Association between reported diet and all-cause mortality: Twenty-one-year follow-up on 27,530 Seventh-day Adventists. *American Journal of Epidemiology* 119:775–787.

Kahuranga, J. 1981. Population estimates, densities and biomass of large herbivores in Simanjiro Plains, Northern Tanzania. *African Journal of Ecology* 19:225–238.

Kandel, W. and E.A. Parrado. 2005. Restructuring of the US meat processing industry and new Hispanic migrant destinations. *Population and Development Review* 31:447–471.

Kaplan, H. et al. 2000. A theory of human life history evolution: Diet, intelligence, and longevity. *Evolutionary Anthropology* 9:156–185.

Karkanas, P. et al. 2007. Evidence for habitual use of fire at the end of the Lower Paleolithic: Site-formation processes at Qesem Cave, Israel. *Journal of Human Evolution* 53:197–212.

Katanoda, K. and Y. Matsumura. 2002. National Nutrition Survey in Japan – Its methodological transition and current findings. *Journal of Nutritional Science and Vitaminology* 48:423–432.

Keenleyside, A. et al. 2006. Stable isotope evidence of diet in a Greek colonial population from the Black Sea. *Journal of Archaeological Science* 33:1205–1215.

Kelleher, K. 2004. *Discards in the World's Marine Fisheries: An Update.* Rome: FAO.

Key, N. and W. McBride. 2007. *The Changing Economics of U.S. Hog Production.* Washington, DC: USDA.

Key, T.J. et al. 2009a. Cancer incidence in vegetarians: Results from the European Prospective Investigation into Cancer and Nutrition (EPIC-Oxford). *American Journal of Clinical Nutrition* 89:1620S–1626S.

Key, T.J. et al. 2009b. Mortality in British vegetarians: Results from the European Prospective Investigation into Cancer and Nutrition (EPIC-Oxford). *American Journal of Clinical Nutrition* 89:1613S–1619S.

Key, N. et al. 2011. *Trends and Developments in Hog Manure Management: 1998–2009.* Washington, DC: USDA.

Keys, A. 1980. *Seven Countries: A Multivariate Analysis of Death and Coronary Heart Disease.* Cambridge, MA: Harvard University Press.

KFC (Kentucky Fried Chicken). 2012. *About KFC.* http://www.kfc.com.my/about-colonel-story.php. Accessed on November 14, 2012.

Kim, S. and B. Dale. 2004. Cumulative energy and global warming impact from the production of biomass for biobased products. *Journal of Industrial Ecology* 7:147–163.

Kingsolver, B. 2002. *Small Wonder.* New York: Harper-Collins.

Kiple, K.F. 2000. The question of Paleolithic nutrition and modern diet. In: K.F. Kiple and K.C. Ornelas, eds., *The Cambridge World History of Food.* Cambridge: Cambridge University Press, pp. 1704–1709.

Klein, R.G. 2009. *The Human Career: Human Biological and Cultural Origins.* Chicago, IL: University of Chicago Press.

Knox, L. and K. Richmond. 2012. *The World Encyclopedia of Meat, Game and Poultry.* London: Anness Publishing.

Koch, P.L. and A.D. Barnosky. 2006. Late Quaternary extinctions: State of the debate. *Annual Review of Ecology, Evolution, and Systematics* 37:215–250.

Koepke, N. and J. Baten. 2005. The biological standard of living in Europe during the last two millennia. *European Review of Economic History* 9:61–95.

Koknaroglu, H., K. Ekinci, and M.P. Hoffman. 2007. Cultural energy analysis of pasturing systems for cattle finishing programs. *Journal of Sustainable Agriculture* 30:5–20.

Konner, M. and S.B. Eaton. 2011. Paleolithic nutrition. *Nutrition in Clinical Practice* 25:594–602.

Kontogianni, M.D. et al. 2008. Relationship between meat intake and the development of acute coronary syndromes: The CARDIO2000 case-control study. *European Journal of Clinical Nutrition* 62:171–177.

Kozlov, A.I. et al. 1998. Gene geography of primary hypolactasia in populations of the Old World. *Genetika* 34:551–561.

Krämer, H.M. 2008. "Not befitting our divine country": Eating meat in Japanese discourses of self and other from the seventeenth century to the present. *Food and Foodways* 16:33–62.

Kucharik, C.J. and N. Ramankutty. 2005. Trends and variability in U.S. corn yields over the twentieth century. *Earth Interactions* 9:1–29.

Kümpel, N.F. et al. 2010. Incentives for hunting: The role of bushmeat in the household economy in rural Equatorial Guinea. *Human Ecology* 38:251–264.

Kuzmin, Y.V. 2009. Extinction of the woolly mammoth (*Mammuthus primigenius*) and woolly rhinoceros (*Coelodonta antiquitatis*) in Eurasia: Review of chronological and environmental issues. *Boreas* 39:247–261.

Landers, T.F. et al. 2012. A review of antibiotic use in food animals: Perspective, policy, and potential. *Public Health Reports* 127:4–22.

Landis, A., S. Miller, and T.L. Theis. 2007. Life cycle of the corn-soybean agroecosystem for biobased production. *Environmental Science & Technology* 41:1457–1464.

Lappé, F.M. 1971. *Diet for a Small Planet.* New York: Ballantine Books.

Lappé, F.M. and A. Lappé. 2002. *Hope's Edge: The Next Diet for a Small Planet.* New York: Tarcher and Penguin.

Larsen, C.S. 2000. Dietary reconstruction and nutritional assessment of past peoples: The bioanthropological record. In: K.F. Kiple and K.C. Ornelas, eds., *The Cambridge World History of Food.* Cambridge: Cambridge University Press, pp. 13–34.

Larsen, C.S. 2003. Animal source foods and human health during evolution. *The Journal of Nutrition* 133:3893S–3897S.

Larsson, S.C. and A. Wolk. 2006. Meat consumption and risk of colorectal cancer: A meta-analysis of prospective studies. *International Journal of Cancer* 119:2657–2663.

Larsson, S.C., J. Virtamo, and A. Wolk. 2011a. Red meat consumption and risk of stroke in Swedish women. *Stroke* 42:324–329.

Larsson, S.C., J. Virtamo, and A. Wolk. 2011b. Red meat consumption and risk of stroke in Swedish men. *American Journal of Clinical Nutrition* 94:417–421.

Lauk, C. and K.-H. Erb. 2009. Biomass consumed in anthropogenic vegetation fires: Global patterns and processes. *Ecological Economics* 69:301–309.

Lawrie, R.A. and D. Ledward, eds. 2006. *Lawrie's Meat Science.* Boca Raton, FL: CRC Press.

Le Bourg, É. 2006. Dietary restriction would probably not increase longevity in human beings and other species able to leave unsuitable environments. *Biogerontology* 7:149–152.

Lee, W.T.K. 2002. Nutrition transition in China – A challenge in the new millennium. *Nutrition & Dietetics* 59:74.

Lee, D.T.S. et al. 2009. Antenatal taboos among Chinese women in Hong Kong. *Midwifery* 25:104–113.

Legge, A.J. and P.A. Rowley-Conwy. 1987. Gazelle killing in Stone Age Syria. *Scientific American* 257(2):88–95.

Leheska, J.M. et al. 2008. Effects of conventional and grass-feeding systems on the nutrient composition of beef. *Journal of Animal Science* 86:3575–3585.

Lehsten, V. et al. 2009. Estimating carbon emissions from African wildfires. *Biogeosciences* 6:349–360.

Leonard, W.R., J.J. Snodgrass, and M.L. Robertson. 2007. Effects of brain evolution on human nutrition and metabolism. *Annual Review of Nutrition* 27:311–327.

Lesschen, J.P. et al. 2011. Greenhouse gas emission profiles of European livestock sectors. *Animal Feed Science and Technology* 166–167:16–28.

Li, K.S. et al. 2004. Genesis of a highly pathogenic and potentially pandemic H5N1 influenza virus in eastern Asia. *Nature* 430:209–213.

Li, D. et al. 2005. Lean meat and heart health. *Asia Pacific Journal of Clinical Nutrition* 14:113–119.

Liebenberg, L. 2006. Persistence hunting by modern hunter-gatherers. *Current Anthropology* 47:1017–1025.

Lipoeto, N.I. et al. 2004. Nutrition transition in west Sumatra, Indonesia. *Asia Pacific Journal of Clinical Nutrition* 13:312–316.

Liu, H. et al. 2009. At-home meat consumption in China: An empirical study. *Agriculture and Resource Economics* 53:485–501.

Lobel, S. 2009. *Lobel's Meat Bible: All You Need to Know About Meat and Poultry.* San Francisco, CA: Chronicle Books.

Lokuruka, M.N.I. 2006. Meat is the meal and status is by meat: Recognition of rank, wealth, and respect through meat in Turkana culture. *Food and Foodways* 14:201–229.

Lovenheim, P. 2002. *Portrait of a Burger as a Young Calf: The True Story of One Man, Two Cows and the Feeding of a Nation.* New York: Harmony Books.

Lowe, M. and G. Gereffi. 2009. *A Value Chain Analysis of the U.S. Beef and Dairy Industries.* http://www.cggc.duke.edu/environment/valuechainanalysis/CGGC_BeefDairyReport_2-16-09.pdf. Accessed on November 14, 2012.

Lu, F. 1998. *Output Data on Animal Products in Chin: How Much They Are Overstated?* Beijing: China Center for Economic Research at Beijing University.

MacDonald, J.M. and W.D. McBride. 2009. *The Transformation of U.S. Livestock Agriculture: Scale, Efficiency, and Risks.* Washington, DC: USDA.

MacDonald, J.M. et al. 2009. *Manure Use for Fertilizer and for Energy.* Washington, DC: USDA.

Maike, D. 2010. Cairo's informal waste collectors. *Erdkunde* 66:27–44.

Mangels, A.R. 2008. Vegetarian diets in pregnancy. In: C.J. Lammi-Keefe, S.C. Couch, and E.H. Philipson, eds., *Nutrition and Health: Handbook of Nutrition and Pregnancy.* Totowa, NJ: Humana Press, pp. 215–231.

Mangels, A.R. and V. Messina. 2001. Considerations in planning vegan diets: Infants. *Journal of American Dietetic Association* 101:670–677.

Manjoo, F. 2012. Fake meat so good it will freak you out. *Slate.* July 26. http://www.slate.com/articles/technology/technology/2012/07/beyond_meat_fake_chicken_that_tastes_so_real_it_will_freak_you_out_.html. Accessed on November 14, 2012.

Mann, N. 2000. Dietary lean red meat and human evolution. *European Journal of Nutrition* 39:71–79.

Mann, N. 2007. Meat in the human diet: An anthropological perspective. *Nutrition & Dietetics* 64:S102–S107.

Marlowe, F. 2010. *The Hadza: Hunter-Gatherers of Tanzania.* Berkeley, CA: University of California Press.

Marshall, B.M. and S.B. Levy. 2011. Food animals and antimicrobials: Impacts on human health. *Clinical Microbiology Reviews* 24:718–733.

Marti, D.L., R.J. Johnson, and K.H. Mathews. 2011. *Where's the (Not) Meat?* Washington, DC: USDA.

Martin, P.S. 1958. Pleistocene ecology and biogeography of North America. *Zoogeography* 151:375–420.

Martin, P.S. 1967. Overkill at Olduvai Gorge. *Nature* 215:212–213.

Martin, P.S. 1990. 40,000 years of extinctions on the "planet of doom." *Palaeogeography, Palaeoclimatology, Palaeoecology* 82:187–201.

Martin, P.S. 2005. *Twilight of the Mammoths*. Berkeley, CA: University of California Press.

Masia, R. et al. 1999. High prevalence of cardiovascular risk factors in Gerona, Spain, a province with low myocardial infarction incidence. *Journal of Epidemiology and Community Health* 52:707–715.

Mason, J. and P. Singer. 1990. *Animal Factories*. New York: Harmon Books.

Masuda, T. and P.D. Goldsmith. 2010. China's meat consumption: An income elasticity analysis and long-term projections.

Matheny, G. and C. Leahy. 2007. Farm-animal welfare legislation, and trade. *Law and Contemporary Problems* 70:325–356.

Matsumura, Y. 2001. Nutrition trends in Japan. *Asia Pacific Journal of Clinical Nutrition* 10(Suppl. 1):S40–S47.

Mattison, J. et al. 2012. Impact of calorie restriction on health and survival in rhesus monkeys from the NIA study. *Nature* 489(7415):318–321.

Mattson, M.P. 2005. Energy intake, meal frequency, and health: A neurobiological perspective. *Annual Review of Nutrition* 25:237–260.

Max Rubner-Institut. 2008. *Nationale Verzehrsstudie II: Ergebnisbericht, Teil 2*. Karslruhe: Max Rubner-Institut. http://www.was-esse-ich.de/. Accessed on November 14, 2012.

McAfee, A.J. et al. 2010. Red meat consumption: An overview of the risks and benefits. *Meat Science* 84:1–13.

McAlpine, C.A. et al. 2009. Increasing world consumption of beef as a driver of regional and global change: A call for policy action based on evidence from Queensland (Australia), Colombia and Brazil. *Global Environmental Change* 19:21–33.

McBride, W.D. and K. Mathews. 2011. *The Diverse Structure and Organization of U.S. Beef Cow-Calf Farms*. Washington, DC: USDA.

McCay, C.M. et al. 1935. The effect of retarded growth upon the length of the life-span and ultimate body size. *Journal of Nutrition* 10:63–79.

McDonald's. 2012. *Better Not Just Bigger*. http://www.mcdonalds.com/us/en/our_story.html. Accessed on November 14, 2012.

McHenry, H.M. and K. Coffing. 2000. *Australopithecus* to *Homo*: Transformations in body and mind. *Annual Review of Anthropology* 29:125–146.

Medeiros, L.C. et al. 2001. *Nutritional Content of Game Meat*. Laramie, WY: University of Wyoming. http://www.huntfish.info/article_uploads/NutritionalContentGame.pdf. Accessed on November 14, 2012.

Mekonnen, M.M. and A.Y. Hoekstra. 2010. *The Green, Blue and Grey Water Footprint of Farm Animals and Animal Products*. Enchede: Twente Water Center, University of Twente.

Menotti, A. et al. 1999. Food intake patterns and 25-year mortality from coronary heart disease: Cross-cultural correlations in the Seven Countries Study. *European Journal of Epidemiology* 15:507–515.

Merrington, G. et al. 2002. *Agricultural Pollution: Environmental Problems and Practical Solutions.* Boca Raton, FL: CRC Press.

Merten, C. et al. 2011. Methodological characteristics of the national dietary surveys carried out in the European Union as included in the European Food Safety Authority (EFSA) Comprehensive European Food Consumption Database. *Food Additives & Contaminants: Part A* 28:975–995.

Meyer-Rochow, V.B. 2009. Food taboos: Their origins and purposes. *Journal of Ethnobiology and Ethnomedicine* 5:1–10.

Micha, R., S.K. Wallace, and D. Mozaffarian. 2010. Red and processed meat consumption and risk of incident coronary heart disease, stroke, and diabetes mellitus. *Circulation* 121:2271–2283.

Micronutrient Initiative. 2009. *United Call to Action on Vitamin and Mineral Deficiencies.* Ottawa, ON, Canada: Micronutrient Initiative. http://www.unitedcalltoaction.org/documents/Investing_in_the_future.pdf. Accessed on November 14, 2012.

Miller, E.R. et al., eds. 1991. *Swine Nutrition.* Boston, MA: Butterworth-Heinemann.

Milligan, K. et al. 1982. Density and biomass of the large herbivore community in Kainji Lake National Park, Nigeria. *African Journal of Ecology* 20:1–12.

Milton, K. 1999. A hypothesis to explain the role of meat-eating in human evolution. *Evolutionary Anthropology* 8:11–21.

Milton, K. 2003. Critical role played by animal source foods in human (*Homo*) evolution. *Journal of Nutrition* 133(Suppl. 2):3886S–3982S.

Monceau, C., É. Blanche-Barbat, and J. Échampe. 2002. De plus en plus de produits élaborés. *INSEE No. 846.* Paris: INSEE.

Morewedge, C.K., Y.E. Huh, and J. Vosgerau. 2010. Thought for food: Imagined consumption reduces actual consumption. *Science* 330:1530–1533.

Morrison, B., P. Eisler, and A. DeBarros. 2009. Old-hen meat fed to pets and schoolkids. *USA Today.* December 16.

Mosimann, J.E. and P.S. Martin. 1975. Simulating overkill by Paleoindians. *American Scientist* 63:304–313.

Murawska, D. et al. 2011. Age-related changes in the percentage content of edible and non-edible components in broiler chickens. *Asian-Australian Journal of Animal Science* 24:532–539.

Muth, M.K. et al. 2006. *Poultry Slaughter and Processing Sector Facility-Level Model.* Research Triangle Park, NC: RTI International.

Muth, M.K. et al. 2011. *Consumer-Level Food Loss Estimates and Their Use in the ERS Loss-Adjusted Food Availability Data.* Washington, DC: USDA. http://www.ers.usda.gov/publications/tb-technical-bulletin/tb1927.aspx. Accessed on November 14, 2012.

Myhrvold, N., C. Young, and M. Bilet. 2011. *Modernistic Cuisine. Volume 3: Animals and Plants.* Bellevue, WA: The Cooking Lab.

NADP (National Atmospheric Deposition Program). 2012. *Ammonium Ion Wet Deposition.* http://nadp.sws.uiuc.edu/maplib/pdf/2010/NH4_dep_10.pdf. Accessed on November 14, 2012.

Nam, K.C., Jo, C., and M. Lee. 2010. Meat products and consumption culture in the East. *Meat Science* 86:95–102.

Navarette, A., C.P. van Schaik, and K. Isler. 2011. Energetics and the evolution of human brain size. *Nature* 480:91–94.

Nawaz, M.S. et al. 2001. Human health impact and regulatory issues involving antimicrobial resistance in the food animal production environment. *Regulatory Research Perspectives* 1:1–10.

Naylor, R. and M. Burke. 2005. Aquaculture and ocean resources: Raising tigers of the sea. *Review of Environment and Resources* 30:185–218.

Naylor, R.L. et al. 2009. Feeding aquaculture in an era of finite resources. *Proceedings of the National Academy of Sciences* 106:15103–15110.

NBS (National Bureau of Statistics). 2001. *China Statistical Yearbook*. Beijing: China Statistics Press.

NCC (National Chicken Council). 2010. *National Chicken Council Animal Welfare Guidelines and Audit Checklist for Broiler Breeders*. Washington, DC: NCC.

NCJDRSU (National CJD Research & Surveillance Unit). 2012. *CJD Statistics*. http://www.cjd.ed.ac.uk/. Accessed on November 19, 2012.

Ndibalema, V.G. and A.N. Songorwa. 2008. Illegal meat hunting in Serengeti: Dynamics in consumption and preferences. *African Journal of Ecology* 46:311–319.

Nepstad, D. et al. 2009. The end of deforestation in the Brazilian Amazon. *Science* 326:1350–1351.

Neumann, C.G. et al. 2003. Animal source foods improve dietary quality, micronutrient status, growth and cognitive function in Kenyan school children: Background, study design and baseline findings. *Journal of Nutrition* 133:3941S–3949S.

Nguyen, T.L.T., J.E. Hermansen, and L. Mogensen. 2010. Fossil energy and GHG saving potentials of pig farming in the EU. *Energy Policy* 38:2561–2571.

NHMRC (National Health and Medical Research Council). 2006. *Nutrient Reference Values for Australia and New Zealand Including Recommended Dietary Intakes*. Canberra: NHMRC.

NIHN (National Institute of Health and Nutrition). 2007. *Outline for the Results of the National Health and Nutrition Survey Japan, 2007*. Tokyo: NIHN.

Njiforti, H.L. 1996. Preferences and present demand for bushmeat in north Cameroon: Some implications for wildlife conservation. *Environmental Conservation* 23:149–155.

Noakes, M., J. Keogh, and P. Clifton. 2007. Obesity and type 2 diabetes mellitus. *Nutrition & Dietetics* 64:S156–S161.

Norat, T. et al. 2002. Meat consumption and colorectal cancer risk: Dose-response meta-analysis of epidemiological studies. *International Journal of Cancer* 98:241–256.

Notenbaert, A. 2009. The role of spatial analysis in livestock research for development. *GIScience & Remote Sensing* 46:1–11.

NRC (National Research Council). 1982. *United States-Canadian Tables of Feed Composition: Nutritional Data for United States and Canadian Feeds, Third Revision*. Washington, DC: NAS.

NRC. 1988. *Nutrient Requirements of Swine.* Washington, DC: National Academy Press.

NRC. 1994. *Nutrient Requirements of Poultry.* Washington, DC: National Academy Press.

NRC. 1996. *Nutrient Requirements of Beef Cattle.* Washington, DC: National Academy Press.

NRC. 2002. *The Scientific Basis for Estimating Air Emissions from Animal Feeding Operations: Interim Report.* Washington, DC: NAS.

Ntiamoa-Baidu, Y. 1998. *Sustainable Harvesting, Production and Use of Bushmeat.* Accra: Ministry of Lands and Forestry.

Nyahongo, J.W. et al. 2005. Benefits and costs of illegal grazing and hunting in the Serengeti ecosystem. *Environmental Conservation* 32:326–332.

Odadi, W.O. et al. 2011. African wild ungulates compete with or facilitate cattle depending on season. *Science* 333:1753–1755.

O'Donnel, K. 2010. *The Meatlover's Meatless Cookbook: Vegetarian Recipes Carnivores Will Devour.* New York: Da Capo.

Oelze, V.M. et al. 2011. Early Neolithic diet and animal husbandry: Stable isotope evidence from three Linearbandkeramik (LBK) sites in Central Germany. *Journal of Archaeological Science* 38:270–279.

Oenema, O. et al. 2009. Integrates assessment of promising measures to decrease nitrogen losses from agriculture in EU-27. *Agriculture, Ecosystems and Environment* 133:280–288.

Ogino, A. et al. 2004. Environmental impacts of the Japanese beef-fattening system with different lengths as evaluated by a life-cycle assessment method. *Journal of Animal Science* 72:2115–2122.

Ogle, M. 2011. *In Beef We Trust: Americans, Meat, and the Making of a Nation.* Boston, MA: Houghton Mifflin Harcourt.

Ogutu, J.O. and H.T. Dublin. 2002. Demography of lions in relation to prey and habitat in the Maasai Mara national Reserve, Kenya. *African Journal of Ecology* 40:120–129.

Ohnuki-Tierney, E. 1993. *Rice as Self: Japanese Identities Through Time.* Princeton, NJ: Princeton University Press.

Oldeman, L.R. 1994. The global extent of land degradation. In: D.J. Greenland and I. Szabolcs, eds., *Land Resilience and Sustainable Land Use.* Wallingford, WA: CABI, pp. 99–118.

Olsen, S.L. 1989. Solutré: A theoretical approach to the reconstruction of Upper Palaeolithic hunting strategies. *Journal of Human Evolution* 18:295–327.

Oomen, G.J.M. et al. 1998. Mixed farming as a way towards a more efficient use of nitrogen in European Union agriculture. *Environmental Pollution* 102:697–704.

Orive, E. et al. 2002. *Nutrients and Eutrophication in Estuaries and Coastal Waters.* New York: Springer.

Osborne, R. 2004. *Greek History.* London: Routledge.

Oths, K.S., A. Carolo, and J.E. Dos Santos. 2005. Social status and food preference in Southern Brazil. *Ecology of Food and Nutrition* 42:303–324.

Otten, J.J., J.P. Hellwig, and L.D. Meyers, eds. 2006. *Dietary Reference Intakes: The Essential Guide to Nutrient Requirements.* Washington, DC: The National Academies Press.

Owen, R. 1861. *Palaeontology, or, a Systematic Study of Extinct Animals and Their Geological Relations.* Edinburgh: Adam and Charles Black.

Pan, A. et al. 2012. Red meat consumption and mortality. *Archives of Internal Medicine* 172:555–563.

Patton, J.Q. 2005. Meat sharing for coalitional support. *Evolution and Human Behavior* 26:137–157.

PCIFAP (Pew Commission on Industrial Farm Animal Production). 2008. *Putting Meat on the Table: Industrial Farm Animal Production in America.* Washington, DC: The Pew Charitable Trust.

Pegues, D.A. and S.I. Miller. 2009. *Salmonella* Species, including *Salmonella typhi.* In: G.L. Mandell, J.E. Bennett, and R. Dolin, eds., *Principles and Practice of Infectious Diseases,* 7th edn. Philadelphia, PA: Elsevier Churchill Livingstone.

Peleg, A.Y. and D.C. Hooper. 2010. Hospital-acquired infections due to gram-negative bacteria. *New England Journal of Medicine* 362:1804–1813.

Pelicano, E.R.L. et al. 2005. Carcass and cut yields and meat qualitative traits of broilers fed diets containing probiotics and prebiotics. *Brazilian Journal of Poultry Science* 7:169–175.

Pellett, P.L. 1990. Protein requirements in humans. *American Journal of Clinical Nutrition* 51:723–737.

Perreault, N. and S. Leeson. 1992. Age-related carcass composition changes in male broiler chickens. *Canadian Journal of Animal Science* 72:919–929.

Perren, R. 1985. The retail and wholesale meat trade 1880–1939. In: D.J. Oddy and D.S. Miller, eds., *Diet and Health in Modern Britain.* London: Croom Helm, pp. 46–65.

PETA (People for Ethical Treatment of Animals). 2008. *Peta Offers $1 Million Reward to First to Make In Vitro Meat.* Norfolk, VA: PETA.

PETA. 2012. *PETA.* http://www.peta.org/. Accessed on November 14, 2012.

Peters, G.M. et al. 2010. Red meat production in Australia: Life cycle assessment and comparison with overseas studies. *Environmental Science & Technology* 44:1327–1332.

Phelan, J.P. and M.R. Rose. 2006. Caloric restriction increases longevity substantially only when the reaction norm is steep. *Biogerontology* 7:161–164.

Pitesky, M.E., K.R. Stackhouse, and F.M. Mitloehner. 2009. Clearing the air: Livestock's contribution to climate change. *Advances in Agronomy* 103:1–40.

Pollan, M. 2006. *The Omnivore's Dilemma: A Natural History of Four Meals.* New York: Penguin Press.

Pöschl, U. et al. 2010. Rainforest aerosols as biogenic nuclei of clouds and precipitation in the Amazon. *Science* 329:1513–1516.

Powell, T.A. et al. 1993. Economics of space allocation for grower-finisher hogs: A simulation approach. *Review of Agricultural Economics* 15:133–141.

Pozio, E. 2007. World distribution of *Trichinella* spp. infections in animals and humans. *Veterinary Parasitology* 149:3–21.

Prairie, Y.T. and C.M. Duarte. 2007. Direct and indirect metabolic CO_2 release by humanity. *Biogeosciences* 4:215–217.

Purcell, N. 2003. The way we used to eat: Diet, community, and history of Rome. *American Journal of Philology* 124:329–358.

Purcell, N. 2011. Cruel intimacies and risky relationships: Accounting for suffering in industrial livestock production. *Society and Animals* 19:59–81.

Pushkina, D. and P. Raia. 2007. Human influence on distribution and extinctions of the late Pleistocene Eurasian megafauna. *Journal of Human Evolution* 54:769–782.

Queensland Government. 2010. *Feedlot Management Terms Explained.* http://www.daff.qld.gov.au/30_7192.htm. Accessed on November 14, 2012.

Quorn. 2012. *Our Products.* http://www.quorn.us/Products/. Accessed on November 14, 2012.

Ragir, S., M. Rosenberg, and P. Tierno. 2000. Gut morphology and the avoidance of carrion among chimpanzees, baboons, and early hominids. *Journal of Anthropological Research* 56:477–512.

Raichlen, S. 2012. *Best Ribs Ever: A Barbecue Bible Cookbook: 100 Killer Recipes.* New York: Workman Publishing Company.

Raj, A.B.M. 2006. Recent developments in stunning and slaughter of poultry. *World's Poultry Science Journal* 26:467–484.

Ramírez, A.R. 2005. *Monitoring Energy Efficiency in the Food Industry.* Utrecht: Universiteit Utrecht.

Rappaport, R.A. 1968. *Pigs for the Ancestors.* New Haven, CT: Yale University Press.

Reijnders, L. and S. Soret. 2003. Quantification of the environmental impact of different dietary protein choices. *American Journal of Clinical Nutrition* 78:664S–668S.

Renaud, S. and M. de Lorgeril. 1992. Wine, alcohol, platelets, and the French paradox for coronary heart disease. *Lancet* 339:1523–1526.

Reumer, J.W.F. 2007. Habitat fragmentation and the extinction of mammoths (*Mammuthus primigenius*, Proboscidea, Mammalia): arguments for a causal relationship. In: R.D. Kahlke, L.C. Maul, and P.P.A. Mazza, eds., *Late Neogene and Quaternary Biodiversity and Evolution: Regional Developments and Interregional Correlations,* Vol. II. Frankfurt: Courier Forschungsinstitut, pp. 279–286.

Revel, J. 1979. Capital city's privileges: Food supply in early-modern Rome. In: R. Foster and O. Ranum, eds., *Food and Drink in History.* Baltimore, MD: Johns Hopkins University Press, pp. 37–49.

Reynolds, C.K., L.A. Crompton, and A.N. Mills. 2011. Improving the efficiency of energy utilisation in cattle. *Animal Production Science* 56:6–12.

Richards, M.P. 2002. A brief review of the archaeological evidence for Paleolithic and Neolithic subsistence. *European Journal of Clinical Nutrition* 56:1270–1278.

Richards, M.P. 2009. Stable isotope evidence for European Upper Paleolithic human diets. In: J.-J. Hublin and M.P. Richards, eds., *The Evolution of Hominin Diets: Integrating Approaches to the Study of Paleolithic Subsistence*. Berlin: Springer Verlag, pp. 252–257.

Richards, M.P. et al. 2003. Stable isotope evidence of diet at Neolithic Çatalhöyük, Turkey. *Journal of Archaeological Science* 30:67–76.

Richards, M.P. et al. 2005. Isotope evidence for the intensive use of marine foods by Late Upper Palaeolithic humans. *Journal of Human Evolution* 49:390–394.

Richerson, P.J., R. Boyd, and R.L. Bettinger. 2001. Was agriculture impossible during the Pleistocene but mandatory during the Holocene? A climate change hypothesis. *American Antiquity* 66:387–411.

Ridoutt, B.G. and S. Pfister. 2010. A revised approach to water footprinting to make transparent the impacts of consumption on global freshwater scarcity. *Global Environmental Change* 20:113–120.

Ridoutt, B.G. et al. 2012. Meat consumption and water scarcity: Beware of generalizations. *Journal of Cleaner Production* 28:127–133.

Rightmire, G.P. 2009. Middle and Later Pleistocene hominins in Africa and Southwest Asia. *Proceedings of the National Academy of Sciences* 106:16046–16050.

Rimas, A. and E.D.G. Fraser. 2008. *Beef: The Untold Story of How Milk, Meat and Muscle Shaped the World*. New York: William Morrow.

Rinehart, K.E. 1996. Environmental challenges as related to animal agriculture – Poultry. In: E.T. Kornegay, ed., *Nutrient Management of Food Animals to Enhance and Protect the Environment*. Boca Raton, FL: Lewis Publishers, pp. 21–28.

Rixson, D. 2010. *The History of Meat Trading*. Thrumpton: Nottingham University Press.

Roberts, R.G. et al. 2001. New ages for the last Australian megafauna: Continent-wide extinction about 46,000 years ago. *Science* 292:1888–1892.

Robinson, T. et al. 2011. *Global Livestock Production Systems*. Rome: FAO.

Rodriguez, L. 2008. *A Global Perspective in the Total Economic Value of Pastoralism: Global Synthesis Report Based on Six Country Valuations*. Nairobi: World Initiative for Sustainable Pastoralism.

Rogers, P.J. and J.E. Blundell. 1990. Physicobiological bases of food choice. *Nutritional Bulletin* 5(Suppl. 1):31–40.

Roques, S. et al. 2011. Idle land for future crop production. *World Agriculture* 2(2):40–42.

de Rossi, P. et al. 2012. *Vegan Cooking for Carnivores: Over 125 Recipes So Tasty You Won't Miss the Meat*. New York: Grand Central Life & Style.

Roth, J.P. 1999. *The Logistics of the Roman Army at War (264 B.C.–A.D. 235)*. Leiden: Brill.

RSPCA (Royal Society for the Prevention of Cruelty to Animals). 2002. *Behind Closed Doors: The Truth About Chicken Bred for Meat.* http://www. agrowebcee.net/fileadmin/user_upload/faw/doc/reports2/RSPCA%20 REPORT%20ON%20WELFARE%20OF%20BROILERS.pdf. Accessed on November 14, 2012.

Rule, D.C. et al. 2002. Comparison of muscle fatty acid profiles and cholesterol concentrations of bison, beef cattle, elk, and chicken. *Journal of Animal Science* 80:1202–1211.

Russell, N. 2002. The wild side of human domestication. *Society and Animals* 10:285–302.

Saatchi, S. et al. 2011. Benchmark map of forest carbon stocks in tropical regions across three continents. *Proceedings of the National Academy of Sciences* 108:9899–9904.

Samman, S. 2007. Zinc. *Nutrition & Dietetics* 64(Suppl. 4):S131–S134.

Sapkota, A.R. et al. 2007. What do we feed to food-producing animals? A review of animal feed ingredients and their potential impacts on human health. *Environmental Health Perspectives* 115:663–668.

Sarwar, G. 1987. Digestibility of protein and bioavailability of amino acids in foods. *World Review of Dietetics* 54:26–70.

Sarwar, G. et al. 1989. Digestibility of protein and amino acids in selected foods as determined by a rat balance method. *Plant Foods and Human Nutrition* 39:23–32.

Savage, W.W. Jr. 1979. *The Cowboy Hero: His Image in American History and Culture.* Norman, OK: University of Oklahoma Press.

Sawyer, G. 1971. *The Agribusiness Poultry Industry: A History of Its Development.* New York: Exposition Press.

Saxena, A.M. 2011. *Vegetarian Imperative.* Baltimore, MD: Johns Hopkins University Press.

Sayers, K. and C.O. Lovejoy. 2008. The chimpanzee has no clothes. *Current Anthropology* 49:87–114.

Schaller, G.B. 1972. *The Serengeti Lion.* Chicago, IL: University of Chicago Press.

Schlink, A.C., M.-L. Nguyen, and G.J. Viljoen. 2010. Water requirements for livestock production: A global perspective. *Revue scientifique et technique de l'Office International des Epizooties* 29:603–619.

Schlosser, E. 2001. *Fast Food Nation: The Dark Side of the All-American Meal.* New York: Harper.

Schroeder, T.V., T.L. Marsh, and J. Mintert. 2000. *Beef Demand Determinants.* Manhattan, KS: Department of Agricultural Economics, Kansas State University.

Scollan, N. 2003. *Strategies for Optimising the Fatty Acid Composition of Beef.* http:// www.aber.ac.uk/en/media/03ch7.pdf. Accessed on November 14, 2012.

Scruton, R. 2006. *Animal Rights and Wrongs.* London: Metro.

Scully, M. 2011. *Dominion: The Power of Man, the Suffering of Animals, and the Call to Mercy.* New York: St. Martin's Press.

Sekikawa, A. et al. 2003. A "natural experiment" in cardiovascular epidemiology in the early 21st century. *Heart* 89:255–257.

Seré, C. and H. Steinfeld. 1996. *World Livestock Production Systems: Current Status, Issues and Trends.* Rome: FAO.

Serramajem, L. et al. 1995. How could changes in diet explain changes in coronary heart disease mortality in Spain – The Spanish Paradox. *American Journal of Clinical Nutrition* 61:S1351–S1359.

Shahin, K.A. and F.A. Elazeem. 2005. Effects of breed and sex and diet and their interactions on carcass composition and tissue weight distribution of broiler chickens. *Archiv Tierzucht* 48:612–626.

Shetty, P.S. 2002. Nutrition transition in India. *Public Health Nutrition* 5(1A):175–182.

Shi, H.T. et al. 2008. Evidence for the massive scale of turtle farming in China. *Oryx* 42:147–150.

Shimshony, A. and M.M. Chaudry. 2005. Slaughter of animals for human consumption. *Revue scientifique et technique de l'Office International des Epizooties* 24:693–710.

Shipman, P.L. 1986. Scavenging or hunting in early hominids: Theoretical framework and tests. *American Anthropologist* 88:27–43.

Shurtleff, W. and A. Aoyagi. 1975. *Book of Tofu.* New York: Ballantine Books.

Shurtleff, W. and A. Aoyagi. 2004. *History of Soybeans and Soyfoods: 1100 B.C. to the 1980s.* Lafayette, CA: Soyinfo Center. http://www.soyinfocenter.com. Accessed on November 14, 2012.

Siegel, E.H. et al. 2005. Growth indices, anemia, and diet independently predict motor milestone acquisition of infants in south central Nepal. *Journal of Nutrition* 135:2840–2844.

Silberbauer, G.B. 1981. *Hunter and Habitat in the Central Kalahari Desert.* Cambridge: Cambridge University Press.

Sillitoe, P. 2002. Always been farmer-foragers? Hunting and gathering in the Papua New Guinea Highlands. *Anthropological Forum* 12:45–76.

da Silva, P.V. et al. 2010. Variability in environmental impacts of Brazilian soybean according to crop production and transport scenarios. *Journal of Environmental Management* 91:1831–1839.

Silva, M. and J.A. Downing. 1995. The allometric scaling of density and body mass: A nonlinear relationship for terrestrial mammals. *The American Naturalist* 145:704–727.

Simoons, F.J. 1979. Question in the sacred-cow controversy. *Current Anthropology* 20:467–476.

Simoons, F.J. 1994. *Eat Not This Flesh: Food Avoidances from Prehistory to the Present.* Madison, WI: The University of Wisconsin Press.

Sims, L.D. et al. 2003. An update on avian influenza in Hong Kong 2002. *Avian Diseases* 47(Suppl. 3):1083–1086.

Sinclair, U. 1906. *The Jungle.* New York: Doubleday.

Sinclair, J. 1999. *Refrigerated Transportation.* Edinburgh: Witherby & Sons.

Singer, P. 1975. *Animal Liberation: A New Ethics for Our Treatment of Animals.* New York: New York Review/Random House.

Sinha, R. et al. 2009. Meat intake and mortality. *Archives of Internal Medicine* 169:562–571.

Sinsin, B. et al. 2002. Abundance and species richness of larger mammals in Pendjari National Park in Benin. *Mammalia* 66:369–380.

Skaggs, J.M. 1973. *The Cattle-Trailing Industry: Between Supply and Demand, 1866–1890.* Lawrence, KS: University Press of Kansas.

Slatta, R.W. 1990. *Cowboys of the Americas.* New Haven, CT: Yale University Press.

SMC(UK). 2012. *Fact Sheet on Schmallenberg Virus.* http://www.sciencemediacentre. co.nz/2012/03/02/smcuk-fact-sheet-on-schmallenberg-virus/. Accessed on November 14, 2012.

Smil, V. 1994. *Energy in World History.* Boulder, CO: Westview Press.

Smil, V. 1999. Crop residues: Agriculture's largest harvest. *BioScience* 49:299–308.

Smil, V. 2001a. *Feeding the World.* Cambridge, MA: The MIT Press.

Smil, V. 2001b. *Enriching the Earth: Fritz Haber, Carl Bosch, and the Transformation of World Food Production.* Cambridge, MA: The MIT Press.

Smil, V. 2002. Worldwide transformation of diets, burdens of meat production and opportunities for novel food proteins. *Enzyme and Microbial Technology* 30:305–311.

Smil, V. 2004. *China's Past China's Future.* London: RutledgeCurzon.

Smil, V. 2005a. Feeding the world: How much more rice do we need? In: K. Toriyama, K.L. Heong, and B. Hardy, eds., *Rice is Life: Scientific Perspectives for the 21st century.* Proceedings of the World Rice Research Conference held in Tokyo and Tsukuba, Japan, November 4–7, 2004. Los Baños: International Rice Research Institute, pp. 21–23.

Smil, V. 2005b. *Transforming the Twentieth Century: Technical Innovations and Their Consequences.* Cambridge, MA: The MIT Press.

Smil, V. 2008. *Energy in Nature and Society: Energetics of Complex Systems.* Cambridge, MA: The MIT Press.

Smil, V. 2011. Harvesting the biosphere: The human impact. *Population and Development Review* 37:613–636.

Smil, V. 2013. *Harvesting the Biosphere.* Cambridge, MA: The MIT Press.

Smil, V. and K. Kobayashi. 2012. *Japan's Dietary Transition and Its Impacts.* Cambridge, MA: The MIT Press.

Smith, W.J. 2010. *A Rat Is a Pig Is a Dog Is a Boy: The Human Cost of the Animal Rights Movement.* New York: Encounter Books.

Smith, D.L. et al. 2002. Animal antibiotic use has an early but important impact on the emergence of antibiotic resistance in human commensal bacteria. *Proceedings of the National Academy of Sciences* 99:6434–6439.

Smith, P.G. and R. Bradley. 2003. Bovine spongiform encephalopathy (BSE) and its epidemiology. *British Medical Bulletin* 66:185–198.

Smith, C.S. and P.J. Urness. 1984. Small mammal abundance on native and improved foothill ranges, Utah. *Journal of Range Management* 37:353–357.

Snacken, R. et al. 1999. The next influenza pandemic: Lessons from Hong Kong, 1997. *Emerging Infectious Diseases* 5:195–203. http://www.cdc.gov/ncidod/eid/vol5no2/snacken.htm. Accessed on November 14, 2012.

Sofos, J.N. 2008. Challenges to meat safety in the 21st century. *Meat Science* 78:3–13.

Sohal, R.S. and R. Weindruch. 1996. Oxidative stress, caloric restriction, and aging. *Science* 273:59–63.

Solway, J.S. and R.B. Lee. 1990. Foragers, genuine or spurious? *Current Anthropology* 31:109–146.

Speakman, J.R. and C. Hambly. 2007. Starving for life: What animal studies can and cannot tell us about the use of caloric restriction to prolong human lifespan. *Journal of Nutrition* 137:1078–1086.

Spencer, C. 2000. The British Isles. In: K.F. Kiple and K.C. Ornelas, eds., *The Cambridge World History of Food*. Cambridge: Cambridge University Press, pp. 1271–1226.

Speth, J.D. and K.A. Spielmann. 1983. Energy source, protein metabolism, and hunter-gatherer subsistence strategies. *Journal of Anthropological Archaeology* 2:1–31.

Stanford, C.B. 1999. *The Hunting Apes: Meat Eating and the Origins of Human Behavior*. Princeton, NJ: Princeton University Press.

Stanford, C.B. and H.T. Bunn, eds. 2001. *Meat-Eating and Human Evolution*. New York: Oxford University Press.

Steinfeld, H. et al. 2006. *Livestock's Long Shadow: Environmental Issues and Options*. Rome: FAO. ftp://ftp.fao.org/docrep/fao/010/a0701e/a0701e00.pdf

Stiner, M.C. 1994. *Honor Among Thieves: A Zooarchaeological Study of Neanderthal Ecology*. Princeton, NJ: Princeton University Press.

Stiner, M.C., R. Barkai, and A. Gopher. 2009. Cooperative hunting and meat sharing 400–200 kya at Qesem cave, Israel. *Proceedings of the National Academy of Sciences* 106:13207–13212.

Stuart, A.J. 2005. The extinction of woolly mammoth (*Mammuthus primigenius*) and straight-tusked elephant (*Palaeoloxodon antiquus*) in Europe. *Quaternary International* 126–128:171–177.

Stuart, T. 2009. *Waste: Uncovering the Global Food Scandal*. London: Penguin Books.

Stuart, A.J. and A.M. Lister. 2007. Patterns of Late Quaternary megafaunal extinctions in Europe and northern Asia. In: R.D. Kahlke, L.C. Maul, and P.P.A. Mazza, eds., *Late Neogene and Quaternary Biodiversity and Evolution: Regional Developments and Interregional Correlations*, Vol. II. Frankfurt: Courier Forschungsinstitut, pp. 287–297.

Subak, S. 1999. Global environmental costs of beef production. *Environmental Economics* 30:79–91.

Subcommittee on Feed Intake. 1987. *Predicting Feed Intake of Food-Producing Animals*. Washington, DC: National Academy Press.

Suttie, J.M., S.G. Reynolds, and C. Batello. 2005. *Grasslands of the World*. Rome: FAO.

Swatland, H.J. 2010. Meat products and consumption culture in the West. *Meat Science* 86:80–85.

Swinnen, J. and P. Squicciarini. 2012. Mixed messages on prices and food security. *Science* 335:405–406.

Takahashi, E. 1984. Secular trend in milk consumption and growth in Japan. *Human Biology* 56:427–437.

Tanaka, J. 1980. *The San Hunter-Gatherers of the Kalahari.* Tokyo: University of Tokyo Press.

Tansey, G. and J. D'Silva, eds. 1999. *The Meat Business: Devouring a Hungry Planet.* London: Earthscan.

Taylor, V.H., M. Misra, and S.D. Mukherjee. 2009. Is red meat intake a risk factor for breast cancer among premenopausal women? *Breast Cancer Research and Treatment* 117:1–8.

Teleki, G. 1973. *The Predatory Behavior of Wild Chimpanzees.* Lewisburg, PA: Bucknell University Press.

Tennie, C., I.C. Gilby, and R. Mundry. 2009. The meat-scrap hypothesis: Small quantities of meat may promote cooperative hunting in wild chimpanzees (*Pan troglodytes*). *Behavioral Ecology and Sociobiology* 63:421–431.

Thaler, A.M. 1999. The United States perspective towards poultry slaughter. In: *Symposium: Recent Advances in Poultry Slaughter Technology.* Champaign, IL: Poultry Science Association, pp. 298–301.

Thane, C.W. and C.J. Bates. 2000. Dietary intakes and nutrient status of vegetarian preschool children from a British national survey. *Journal of Human Nutrition and Dietetics* 13:149–162.

Thaxton, J.P. et al. 2006. Stocking density and physiological adaptive responses of broilers. *Poultry Science* 85:344–351.

The Vegetarian Resource Group. 2001. How many teens are vegetarian? How many kids don't eat meat? *Vegetarian Journal.* www.vrg.org/journal/vj2001jan/2001janteen.htm. Accessed on November 14, 2012.

Thevenot, R. 1979. *A History of Refrigeration Throughout the World.* Paris: International Institute of Refrigeration.

Thieme, H. 1997. Lower Paleolithic hunting spears from Germany. *Nature* 385:807–810.

Thirsk, J. 2006. *Food in Early Modern England: Phases, Fads, Fashions, 1500–1760.* London: Hambledon Continuum.

Thomas, M. et al. 2008. DNA from pre-Clovis human coprolites in Oregon, North America. *Science* 230:786–789.

Thomsen, P.M., ed. 2011. *The U.S. EU Beef Hormone and Poultry Disputes.* New York: Nova Science Publishers.

Tivall. 2012. *Products.* http://www.tivall.co.uk/product.asp?id=32. Accessed on November 19, 2012.

Toshima, H., Y. Koga, and H. Blackburn, eds. 1994. *Lessons for Science from the Seven Countries Study.* Tokyo: Springer.

Toutain, J.-C. 1971. *La consommation alimentaire en France de 1789 à 1964. Economies et Sociétés.* Geneva: Droz.

Trinkhaus, E. 2005. Early modern humans. *Annual Review of Anthropology* 34:207–230.

Troy, C.S. et al. 2001. Genetic evidence for Near-Eastern origins of European cattle. *Nature* 410:1088–1091.

Truswell, A.S. 2007. Vitamin B12. *Nutrition & Dietetics* 64(Suppl. 4):S120–S125.

Tsatsarelis, C.A. and D.S. Koundouras. 1994. Energetics of baled alfalfa production in northern Greece. *Agriculture, Ecosystems and Environment* 49:123–130.

TsSU (Tsentral'noie statisticheskoie upravlenie). 1977. *Narodnoie khoziaistvo SSSR za 60 let.* Moscow: Statistika.

Turner, J., L. Garcés, and W. Smith. 2005. *The Welfare of Broiler Chickens in the European Union.* Petersfield: Compassion in World Farming Trust. http://www.ciwf.org.uk/includes/documents/cm_docs/2008/w/welfare_of_broilers_in_the_eu_2005.pdf. Accessed on November 14, 2012.

Turzi, M. 2012. Grown in the cone: South America's soybean boom. *Current History* 111(742):5055.

UCS (Union of Concerned Scientists). 2001. *Hogging It: Estimates of Antimicrobial Use in Livestock.* Washington, DC: UCS.

UNFCCC (United Nations Framework Convention on Climate Change). 2012. *Global Warming Potentials.* http://unfccc.int/ghg_data/items/3825.php. Accessed on November 14, 2012.

Unger, T.A. 1997. *Pesticide Synthesis Handbook.* Norwich, NY: William Andrew Publishing.

USBC (US Bureau of the Census). 1975. *Historical Statistics of the United States.* Washington, DC: USBC.

USDA (United States Department of Agriculture). 1992. *Weights, Measures, and Conversion Factors for Agricultural Commodities and Their Products.* Washington, DC: USDA. http://webarchives.cdlib.org/sw1s17tt5t/http://ers.usda.gov/publications/ah697/ah697.pdf. Accessed on November 14, 2012.

USDA. 1995. *Poultry and Eggs.* http://www.aphis.usda.gov/animal_health/emergingissues/downloads/1poultry.pdf. Accessed on November 14, 2012.

USDA. 1999. *Generic HACCP Model for Poultry Slaughter.* Washington, DC: USDA.

USDA. 2010a. *Dietary Guidelines for Americans 2010.* Washington, DC: US DHHS. http://health.gov/dietaryguidelines/dga2010/DietaryGuidelines2010.pdf. Accessed on November 14, 2012.

USDA. 2010b. *USDA Announce New Performance Standards for Salmonella and Campylobacter.* Washington, DC: USDA.

USDA. 2011. *National Nutrient Database for Standard Reference Release 24.* Washington, DC: USDA. http://ndb.nal.usda.gov/ndb/foods/list. Accessed on November 14, 2012.

USDA. 2012a. *Food Availability (Per Capita) Data System.* Washington, DC: USDA. http://www.ers.usda.gov/data-products/food-availability-(per-capita)-data-system.aspx. Accessed on November 14, 2012.

USDA, 2012b. *Food Expenditures.* http://www.ers.usda.gov/data-products/food-expenditures.aspx. Accessed on November 14, 2012.

USDA. 2012c. *Fertilizer Use and Price.* http://www.ers.usda.gov/data-products/fertilizer-use-and-price.aspx. Accessed on November 14, 2012.

Valiathan, M.S. 2003. *The Legacy of Caraka.* London: Orient Longman.

Vanderhaeghen, W. et al. 2010. Methicillin-resistant *Staphylocossus aureus* (MRSA) in food production animals. *Epidemiology and Infection* 138:606–625.

Van der Kooi, M.E. 2010. The inconsistent vegetarian. *Society and Animals* 18:291–305.

Van Winckel, M. et al. 2011. Vegetarian infant and child nutrition. *European Journal of Pediatrics* 170:1489–1494.

Vartanyan, S.L. et al. 1995. Radiocarbon dating evidence for mammoths on Wrangel Island, Arctic Ocean, until 2000 BC. *Radiocarbon* 37:1–6.

Verge, X.P. et al. 2008. Greenhouse gas emissions from the Canadian beef industry. *Agricultural Systems* 98:126–134.

Villa, P. et al. 2005. New data from Ambrona: Closing the hunting versus scavenging debate. *Quaternary International* 126–128:223–250.

Voeten, M.M. and H.H.T. Prins. 1999. Resource partitioning between sympatric wild and domestic herbivores in the Tarangire region of Tanzania. *Oecologia* 120:287–294.

Volatier, J.L. et al. 2007. Contribution of meat and meat products to nutrient intakes of the French adult population. *Sciences des Aliments* 27:123–132.

de Vos, A. 1977. Game as food. *Unasylva* 29(111):2–12.

de Vries, M. and I.J.M. de Boer. 2010. Comparing environmental impacts for livestock products: A review of life cycle assessments. *Livestock Science* 128:1–11.

Wadsworth, G. 1984. *The Diet and Health of Isolated Populations.* Boca Raton, FL: CRC Press.

Waldau, P. 2011. *Animal Rights: What Everyone Needs to Know.* Oxford: Oxford University Press.

Warriss, P.D. 1984. Exsanguination of animals at slaughter and the residual blood content of meat. *Veterinary Record* 115:292–295.

Watanabe, Z. 2008. The meat-eating culture of Japan at the beginning of westernization. *Food Culture* 9:2–8. http://kiifc.kikkoman.co.jp/foodculture/pdf_09/e_002_008.pdf. Accessed on November 14, 2012.

Watanabe, Y. and N. Suzuki. 2006. Is Japan's milk consumption saturated? *Journal of Faculty of Agriculture, Kyushu University* 51:165–171.

Waters, A.E. et al. 2011. Multidrug-resistant *Staphylococcus aureus* in US meat and poultry. *Clinical Infectious Diseases* 52:1227–1230.

WCRF (World Cancer Research Fund) and AICR (American Institute for Cancer Research). 2009. *Policy and Action for Cancer Prevention.* Washington, DC: WCRF and AICR.

Webb, S. 2008. Megafauna demography and late Quaternary climatic change in Australia: A predisposition to extinction. *Boreas* 37:329–345.

Weber, C.L. and H.S. Matthews. 2008. Food-miles and relative climate impacts of food choices in the United States *Environmental Science & Technology* 42:3508–3513.

Webster, A.B., D.L. Fletcher, and S.I. Savage. 1996. Humane on-farm killing of spent hens. *Applied Poultry Science* 5:191–200.

Weidema, B. et al. 2008. Carbon footprint: A catalyst for life cycle assessment. *Journal of Industrial Ecology* 12:3–6.

Weindruch, R. and R.L. Walford. 1988. *The Retardation of Aging and Disease by Dietary Restriction.* Springfield, IL: Charles C. Thomas.

Wells, H.F. and J.C. Buzby. 2008. *Dietary Assessment of Major Trends in U.S. Food Consumption, 1970–2005.* Washington, DC: USDA.

Weng, X. and B. Caballero. 2007. *Obesity and Its Related Diseases in China: The Impact of the Nutrition Transition in Urban and Rural Adults.* Youngstown, NY: Cambria Press.

White, J.H. Jr. 1993. *The American Railroad Freight Car.* Baltimore, MD: The Johns Hopkins University Press.

Whiten, A. and E.M. Widdowson, eds. 1992. *The Human Origins Program and Evolutionary Ecology in Anthropology Today. Foraging Strategies of Monkeys, Apes, and Humans.* Oxford: Clarendon Press.

Whittemore, C.T. 1993. *The Science and Practice of Pork Production.* Essex: Longmann, Essex.

WHO (World Health Organization). 2007. *Nitrate and Nitrite in Drinking Water.* Geneva: WHO.

Wijkström, U.N. 2009. The use of wild fish as aquaculture feed and its effects on income and food for the poor and the undernourished. In: M.R. Hasanand and M. Halwart, eds., *Fish as Feed Inputs for Aquaculture: Practices Paper. No. 518.* Rome: FAO, pp. 371–407.

Wilcox, B.J. et al. 2007. Caloric restriction, the traditional Okinawan diet, and healthy aging. *Annals of the New York Academy of Sciences* 1114:434–455.

Wilkins, J.M. and S. Hill. 2006. *Food in the Ancient World.* Malden, MA: Blackwell Publishing Ltd.

Wilkinson, J.M. 2011. Re-defining efficiency of feed use by livestock. *Animal* 5:1014–1022.

Williams, P. 2007. Nutritional composition of red meat. *Nutrition & Dietetics* 64(Suppl. 4):S113–S119.

Williamson, C.S. et al. 2005. Red meat in the diet. *British Nutrition Foundation, Nutrition Bulletin* 30:323–355.

Wint, G.R.W. and T.P. Robinson. 2007. *Gridded Livestock of the World 2007.* Rome: FAO.

Woolgar, C. 2001. Fast and feast: Conspicuous consumption and the diet of the nobility in the fifteenth century. In: M. Hicks, ed., *Revolution and Consumption in Late Medieval England.* Woodbridge: Boydell, pp. 7–25.

Worsley, A. and G. Skrzypiec. 1997. Teenage vegetarianism: Beauty or the beast? *Nutrition Research* 17:391–404.

Wrangham, R. 2009. *Catching Fire: How Cooking Made Us Human.* New York: Basic Books.

WRAP (Waste & Resources Action Programme). 2008. *The Food We Waste.* http://www.ns.is/ns/upload/files/pdf-skrar/matarskyrsla1.pdf. Accessed on November 19, 2010

Wulf, D.M. 2012. *Did the Locker Plant Steal Some of My Meat?* http://www.caes.uga.edu/Topics/sustainag/documents/didthelockerstealmymeat.pdf. Accessed on November 14, 2012.

Xue, X. and A.E. Landis. 2010. Eutrophication potential of food consumption patterns. *Environmental Science & Technology* 44:6450–6456.

Yang, B.W. et al. 2011. Prevalence of Salmonella on raw poultry at retail markets in China. *Journal of Food Protection* 74:1724–1728.

Ye, Y. et al. 2012. Rebuilding global fisheries: The World Summit goal, costs and benefits. *Fish and Fisheries.*

Young, L.L. et al. 2001. Effects of age, sex, and duration of postmortem aging on percentage yield of parts from broiler chicken carcasses. *Poultry Science* 80:376–379.

Your Guide to Pizza. 2012. *Pizza Trivia and Facts.* http://www.yourguidetopizza.com/TheWorldofPizza/PizzaTrivia.aspx. Accessed on November 14, 2012.

Yule, J.V. 2009. North American late Pleistocene megafaunal extinctions: Overkill, climate change, or both? *Evolutionary Anthropology* 18:159–160.

Zeder, M.A. 2008. Domestication and early agriculture in the Mediterranean Basin: Origins, diffusion, and impact. *Proceedings of National Academy of Sciences* 105:11597–11604.

Zhao, S. et al. 2012. Comparison of the prevalences and antimicrobial resistances of Escherichia coli isolates from different retail meats in the United States, 2002 to 2008. *Applied Environmental Microbiology* 78:1701–1707.

Zohary, D. and M. Hopf. 2000. *Domestication of Plants in the Old World.* Oxford: Oxford University Press.

Index

Acids
  amino, 17–18, 129
  fatty, 9–10, 14–17, 42, 47, 185, 192
    monounsaturated (MUFA), 14–15
    polyunsaturated (PUFA), 14–16, 192
    saturated (SFA), 14–15
Africa, 9, 35–6, 39, 43, 56, 67, 74, 80, 87–8, 114, 121–2, 134, 152, 155, 181, 194, 200, 203–4, 206, 214
  pastoralism in, 117, 119, 151, 153–4
  wild meat in, 34, 50–51, 55
Alfalfa, 114, 131, 134, 156–9, 164, 207
Alpaca, 54
America
  North, 11, 26, 33–4, 42–3, 48, 50, 64, 71, 73, 76, 78–9, 89, 105, 110, 117, 120–121, 170, 179, 209
  South, 55, 66, 71, 80, 87, 114, 117, 119, 121, 123, 127, 151–3, 161, 174, 181, 200, 202–3, 206
Ammonia, 75, 79, 146, 158, 166–9, 197, 208
Animals
  domesticated, 3, 6, 8, 12–13, 29, 32–4, 42, 45, 47, 49–56, 59, 63, 68, 75, 91, 117, 139, 141, 143–4, 147–9, 154, 160, 170, 173, 177
  draft, 14, 52, 54–5, 56–8, 66, 73, 75, 114, 118, 120, 127
  exports of, 79–80
  and greenhouse gas emissions, 168–75
  land requirements of, 147–51
  liberation of, 94, 142–4
  life cycles of, 91–4
  slaughter of, 5, 8, 25, 28, 45, 55, 58, 61, 64, 72, 76, 85, 87, 89–98, 108–9, 111, 118–19, 121, 123, 132, 137–45, 149, 154–5, 179
  stocking densities of, 51, 122, 125–6, 149, 150, 154, 204

Should We Eat Meat?: Evolution and Consequences of Modern Carnivory,
First Edition. Vaclav Smil.
© 2013 John Wiley & Sons, Ltd. Published 2013 by John Wiley & Sons, Ltd.